GLOBAL ENVIRONMENTAL ECONOMICS

GLOBAL ENVIRONMENTAL ECONOMICS

by

HANS W. GOTTINGER

International Institute for Environmental
Economics and Management (IIEEM)
and Nagasaki University

SPRINGER-SCIENCE+BUSINESS MEDIA, B.V.

A C.I.P. Catalogue record for this book is available from the Library of Congress.

ISBN 978-1-4613-7482-4   ISBN 978-1-4615-5435-6 (eBook)
DOI 10.1007/978-1-4615-5435-6

Printed on acid-free paper

CONTENTS

Preface

The movement of carbon from sources to final disposition is known as the carbon cycle. The largest reservoir of carbon is in carbonate sediments such as limestone and chalk. Other significant but less stable reservoirs include fossil fuels, living and dead plants and animals, carbonates and bicarbonates dissolved in the ocean. Huge quantities of carbon, in the form of carbon dioxide (CO_2), move from the biosphere and oceans to the atmosphere and back each year. It is widely believed that before the industrial revolution these flows were closely balanced and the atmospheric CO_2 level was relatively stable. In more recent times, though, several changes in the carbon cycle may have occurred. While it is not disputed that the increase of CO_2 will have an impact on warming and climatic change, uncertainty about fossil energy use, land use change, the carbon cycle, and other climatic disturbances all make prediction of the future climate uncertain. Carbon dioxide in the atmosphere affects the radiation balance of the earth and increasing CO_2 concentrations are expected to cause a warmer climate. Carbon dioxide is relatively transparent to energy as sunlight, but reflects or traps a large portion of this energy when radiated from the earth as heat. A very important development in climate research is the identification of other trace or greenhouse gases (GHG) with this same heat trapping property. The prediction of future temperature is further complicated by diverse factors. Significant feedback effects are expected to accompany any direct effects of CO_2 heat trapping. Warming may change cloud, snow, and ice cover and alter the earth's "albedo" or brightness. Consequently, the reflection and absorption of energy may increase or decrease. Another major concern is that a warmer earth would result in more rapid and violent fluctuations in weather patterns resulting in mounting damages and related possible problems include changes in the range of crop pests, greater health problems from the spread of diseases associated with tropical climates, and disruption of world fisheries. The concern about global climate change as a result of the build up of excessive concentrations of greenhouse gases is a principal concern in environmental policy. The concentration of greenhouse gases in the atmosphere is a stock pollutant, it can take hundreds of years before they lose their properties or disappear. Following climate projections and updates by the Intergovernmental Panel on Climate Change (IPCC), the expected trends in worldwide energy use are such that these could generate important climate changes. The time frame and the magnitude of damages are still highly uncertain. From the economic point of view there are some interesting aspects with the overall problem. Greenhouse gases are uniformly mixed stock pollutants and the problem is on a worldwide scale.

On this background of climate change induced by the accumulation of greenhouse gases we apply an instrumental framework of economic analysis that is conducive to provide an integrated view of the issues. Global Environmental Economics therefore is an integrated economic assessment of GHG emissions impacting economic growth and welfare change along major considerations of uncertainty, technological change and cooperation. I emphasise a theoretical treatment of the issues, using modern tools of economic theory and based on "simple" models that yield illustrative results in terms of "what-if" questions. With this focus the book distinguishes itself uniquely from other economic analytic approaches in the field that have appeared in the past few years.

As such these models could serve as a selective device for much more powerful large-scale, numerically supported models or even form the core of model-based decision support systems. After all, I concur with Robert Lucas who said that "useful policy discussions are ultimately based on models".

In addition to providing a framework for research in global environmental economics, being accessible to a wide array of disciplines focusing on global change, I have also attempted to design a structured textbook on global environmental economics as a course for advanced undergraduates and beginning graduates in economics and social science. This book was started during my tenure at Oxford Institute for Energy Studies and Nuffield College, Oxford, from 1990 to 1992, and, with interruptions, work was further carried on when visiting CICERO, University of Oslo, 1993 to 1994, and the University of Nagasaki, Japan, from 1996. I am very grateful to these institutions for providing research and financial support as well as a stimulating intellectual environment. Subsequently, materials from the book have been presented at various seminars at Oxford University and the University of Oslo, the Stockholm Environment Institute, the University of Montreal, the University of Lisbon, Hong Kong University of Science and Technology, Kyoto University and UN WIDER, Helsinki. I was fortunate enough to receive valuable advice and comments from Professors Terence Gorman and Christopher Bliss, Oxford, Bernt Stigum and Michael Hoel, Oslo, Arnold Zellner, Chicago, Leonid Hurwicz, Minneapolis, Roy Radner, New York and

Hirofumi Uzawa, Tokyo, to all of them I am very much indebted and grateful. I also appreciate the comments by some anonymous referees.

Over all those years, through many changes at draft stages, Ms Victoria Harper did the entire manuscript without ever losing her cheerful temper. Thanks Vicki. I am also grateful to Ms Claudia Witte for support at various stages.

September 1, 1997
Hans W. Gottinger

Introduction

The major purpose of this book is to contribute to better economic policy analysis through improvements in models studying the economic impacts of carbon dioxide (CO_2) and other greenhouse gases, and to show ways in which economic instruments can effectively be put to use to alleviate such problems.

This approach differs in at least one major respect from common studies of the climatic change problem. We focus on the analysis, control and optimization of modelling forms rather than the collection and analysis of data. More concretely, we search for optimal fossil fuel use, research and technology policies rather than prediction of the future. Most studies of the problem exogenously specify technical developments and fossil fuel control policies and then predict future climatic changes. These prediction models incorporate a great deal of data and tend to be quite complex.

The advantage of an optimizing control model is increased flexibility in structural and dynamic assumptions on the economy allowing explicit 'what-if' questions to be asked about the possibility of controlling the growth in atmospheric CO_2 concentrations. We are able to use these models to draw some specific policy conclusions. By using and improving such models we intend to explore a wide range of substantial issues of CO_2 emission analysis of the subject. For example, questions such as;

- are different import/export taxes on fossil fuels desirable in competitive and non-competitive world markets?

- do increasingly severe economic impacts of CO_2 reduce present optimal fossil fuel use?

- do different relations between output and energy use have corresponding impacts on the long-run optimal level of atmospheric CO_2?

- in which cases can a paradoxical situation arise where a lack of cooperation among nations in controlling CO_2 leads to lower initial world fossil fuel use?

- what are the impacts of change in important parameters, such as the discount rate, risk aversion, energy productivity, and CO_2 retention on the present level of fossil fuel use and the long-run CO_2 level?

- what are the conditions under which the optimal use of fossil energy falls or rises during the transition to a future steady state?

Because some present energy-economy-environmental (EEE) models do not optimize and do not treat such issues as cooperation, technical change and uncertainty in an integrated manner, these models may be too pessimistic about the possibility of controlling the growth in atmospheric CO_2 concentrations.

The omissions and the structure of present models place hidden, rigid and unnecessary constraints on the economy. We hope that this book helps to eliminate some of these problems and allow a more solid foundation from which policy conclusions can be drawn.

The global nature of potential CO_2 problems makes it very difficult to keep models simple and makes issues of uncertainty and international cooperation very important. The long time span lends significance to the modelling of technical change and also stresses the need to balance the welfare of present and future generations.

Previous and Related Studies

Doing research in economic models and applying them to the CO_2 problem requires that we take the physical aspects and findings of this problem as given inputs. We see that although controversy and uncertainty surround almost all the specific predictions of CO_2 effects, the basic conclusion that a doubling of atmospheric CO_2 will cause a significant temperature rise has remained constant. Variations in the physical parameters, based on different sources, can be confidently handled by sensitivity analysis in the numerical treatment of our models. An important more recent development is the emphasis on the additional temperature increases due to trace gases or so-called greenhouse gases (GHG), to which we will respond in our models. Since CO_2 impact assessment would also result in a CO_2 equivalent change in GHG emissions our analysis sometimes uses the terms interchangeably.

Before we clearly identify the relevant work in this area we have some general remarks on some energy-economy studies used to examine CO_2 and other related energy issues. We identify several limitations in these models which we will address in our models. We find that most major studies of the CO_2 problem use predictive models which do not include feedback effects. In the models we discuss, the process of accumulating and disposing of physical capital is handled by a variety of ad hoc methods and is not modelled explicitly. Technical progress is often considered to occur uniformly throughout the economy or is included by predicting specific technical developments. Though the treatment of technical progress is not deep in most studies of CO_2, the importance of predicting technical progress is recognized. Finally, the relationship between trade patterns and CO_2 accumulation is seldom considered in present energy - economy models.

Two major kinds of economic studies can be identified for dealing with the CO_2 problem. The first category treats economic and economic modelling issues in the context of an integrated framework of energy-economy and climate changes, the second category applies the theory of resource use and depletion to the management of CO_2 emissions.

In what follows we provide a selected overview of those studies which yield pertinent results comparable to our own or which are of methodological interest in modelling the energy-economy-climate interactions.

Some years ago the US National Research Council (1983) compiled a detailed investigation that up to now constitutes the most extensive, comprehensive and consistent examination of the climate change problem. It uses energy-economy, climate and agricultural models to predict future impacts of carbon dioxide and trace gas accumulation. The major conclusions are that no radical actions should be taken, that increases in carbon dioxide are likely, and that more research is necessary. In this report, the developers of the energy-economy model (W. Nordhaus and C. Yohe) note that the technology development and elasticity of substitution parameters critically affect the model's results.

It should be added that the method of modelling technical change and energy substitution possibilities is also critical and controversial. The economic modelling chapters of the NRC report have been updated by Yohe (1984). The conclusions of the report have not changed significantly.

Another collaborative study, the joint MIT-Stanford study (Rose, Miller, and Agnew, 1983), is of interest because it is one of few reports which search for alternatives to increasing CO_2 and offer some positive choices. In the Edmonds and Reilly model (Edmonds and Reilly, 1983) S-shaped paths are exogenously specified for several new energy technologies. The MIT-Stanford study modifies these paths and also looks at additional technologies. It finds that the adoption of realistic CO_2 reducing technologies, while not eliminating a significant CO_2 warming could increase the CO_2 doubling time to several centuries.

In attempting to discuss optimizing strategies W. Nordhaus (1980) made a seminal contribution by applying simple optimization models to the qualitative and quantitative analysis of the CO_2 problem. By letting the consumption equation depend on fossil energy use he determines the appropriate tax policy to control CO_2 and makes a quantitative estimate of this tax.

In most present studies of the CO_2 problem we find that technological change and technology substitution are specified exogenously or modelled in a very simple fashion.

In this regard, the logistic (S-curve) assumption on the diffusion of new technologies has become very popular, although it lacks sufficient economic explanatory power. For example, in a study by Perry et al (1982) an energy demand level and a fossil fuel use pattern is assumed. Fossil fuel use follows a logistic curve between the present and an assumed ultimate level of use. The rate of non-fossil energy growth needed to fill the gap, between fossil energy use and assumed total energy demand is then examined. This study emphasizes the importance of analysing the investment needed in non fossil energy to fill the gap but it does not present a model of the substitution process. In the model we suggest, the substitution process is a direct result of our maximization of welfare.

Modelling the impacts of changes in energy use on the economy is a major problem. A good starting point for such considerations would be the ETA-Macro model (Manne, 1977) though it has not been used for studying the CO_2 problem. This model can be described as a multisector, forward-looking model. It examines consumption and investment policies and their impact on national welfare. National welfare is measured by discounting utility from the present to a distant horizon. ETA-Macro consists of two models: a macro-model of the whole economy and a more detailed model of the energy sector. The model seems more sophisticated in its treatment of capital and the determination of the desirable level of energy use than those economic models presently used in CO_2 analysis. However, it is limited to the USA in geographic scope which makes it unsuitable for examining international problems such as CO_2. The model has no endogenous technical progress which we consider an important feature for the analysis of CO_2. On the other hand, the model has several features which would be desirable in models of energy, economy and the environment. It is optimizing and considers costs and benefits of capital investment. More recently, Manne (1992a) has proposed Global 2100, a model adapted to the CO_2 problem but with similar structural features of ETA-Macro. The more recent study by Nordhaus (1994) centers around the construction, integration, model assessment and policy analysis of his economic control model DICE (Dynamic Integrated Model of Climate and Economy). DICE is a dynamic, intertemporal, optimal, interactive, welfare-economic control model based on structural equation constraints such as population growth, production constraints, capital stock accumulation, emission constraints (where GHGs are normalized by their carbon dioxide equivalent in terms of their 'global warming potential' (GWP)). The model contains a critical economy-climate interface that links GHG emissions to their accumulation and transport in the atmosphere, the radiative forcing of the GHGs and their links to climate change. To assess the economic impacts such as damages, DICE contains feedbacks from climate change to economics, by specifying the loss of global output due to climate change. The climate part of DICE relates to specifications of General Circulation Models (GCMs), condensed as a 'minimodel' of climate change to have it fit with the economic interface. Given the structure of DICE, Nordhaus puts his model to test. First he estimates damage profiles of GHG induced damages, for particular sectors as well as enticing the entire GDP loss. Furthermore, he looks at the welfare economic implications (net benefits) of seven major policy strategies to control global climatic change: (1) no controls, (2) optimal policy, (3) ten-year delay of optimal policy, (4) stabilising emissions at 1990 rates, (5) 20 percent emission reduction from 1990 levels, (6) geoengineering, (7) climate stabilisation with upper limit of total mean temperature increase by 1.5C from 1990.

The extent of uncertainty in the model parameters gives rise to estimating the impact range on strategic outcomes as well as it applies to regulatory decision-making on how to optimally impose regulatory controls to minimize over- or undershooting of environmental regulation and policy measures (the value of information of waiting vs acting). Choosing the level of GHG emission-limiting regulations that will maximize social welfare by optimally balancing the costs of emission control against the benefit of decreased environmental damage is inherently not possible, because of pervasive uncertainty about the likely size of the critical GHG budget, its relationship to the quantity of GHG emitted, the effects of GHG in the atmosphere, and the appropriate valuation of these consequences.

In setting up our approach we were influenced by applications of optimal control models to pollution problems (Fisher, 1981; Conrad and Clark, 1987) and relevant issues of resource use. Such models often show different structures, e.g. pollution affecting utility or production, the pollutant acting as a stock or a flow and the way abatement activities are available.

In this specific context, we find that many models and results on the depletion of a non-renewable resource could be applied with simple modifications to the CO_2 problem. Under two assumptions the problem of fossil fuel use in the face of increasing carbon dioxide is parallel to the problem of consumption of a limited resource. The first assumption is that the carbon dioxide rate is sufficiently small to be ignored. The second is that CO_2 impacts follow a 'step' pattern: that is, CO_2 has no impact on productivity until a critical level, M_c, is reached. Then if the CO_2 level exceed M_c production falls to zero, or remains stagnant. One of the most serious effects in facing global climatic change is that of irreversibility, that is, given the accumulation of atmospheric CO_2 we will reach a critical level of the CO_2 budget where there is a point of no return (unless technologies are in place that effectively remove CO_2 from the budget). The interpretation of a critical level of atmospheric CO_2 accumulation where there is a precipitous drop of production means that we have reached the biophysical limits of growth.

An interesting treatment of endogenous neutral technical progress in a depletion model was suggested by Chiarella (1980). He proves the existence of a steady state growth path and a simple rule governing the rate of investment in research. Research investment along the optimal path should be carried out until the growth rate in the marginal accumulation of technology equals the difference between the marginal product due to an extra unit of research investment and the marginal product of capital.

A similar problem is the use of a limited non-renewable resource when the reserve of the resource is unknown. The model by Gilbert (1979) can be directly converted to a model of fossil fuel when the critical CO_2 level is uncertain. Under the above assumptions this problem is equivalent to determining the rate of fossil fuel use when the critical concentration of atmospheric carbon dioxide is unknown. The results show that the optimal use of fossil fuel is lower when uncertainty is properly considered than when the expected values are assumed to be certain.

Deshmukh and Pliska (1980, 1983) study more complex models of the same problem. The possibility of doing exploration to find new reserves is a significant addition in their models. The parallel in the CO_2 problem is research to increase the probability of finding a technology for the removal of CO_2 from the atmosphere. Their findings imply that in the periods between discoveries or research breakthroughs, fossil fuel use and consumption fall, but if research is very successful long-run fuel use may rise.

Conceptual Framework: The Reference Model

In exploring the shifts in energy, economy and CO_2 interactions we start with a basic 'simple' model assuming that energy is the only productive factor in the economy and that energy use causes CO_2 accumulation which lowers productivity. Consumption of goods in the economy causes a flow of utility. The decision-maker acts to maximise the integral of the discounted utility stream. The model has a single state variable, CO_2 concentration, and is autonomous. Unique features of the model are the very general production function and the assumption that the pollution does not affect utility directly.

Special interesting features of this model include:
- conditions which assure the existence of an optimal equilibrium,
- a phase plane analysis showing the conditions under which the shadow price of fossil fuels rises, and what conditions assure convergence to the equilibrium
- definition of an ideal tax to control CO_2 in a decentralized economy and factors which make estimating the tax difficult.

We develop models which can be solved numerically and applied to specific issues of interest. Two specific forms are presented. In the first, the negative impacts of CO_2 accumulation occur abruptly at specific levels of atmospheric CO_2; this is referred to as the step model of CO_2 impacts. In the second form, the negative impacts of CO_2 increase gradually and continuously with increases in atmospheric CO_2. We refer to this model as the modified Cobb-Douglas model. Furthermore, on the basis of this reference model, we will develop and assess several other useful economic models for CO_2 studies as well as supporting tools and models.

One model in particular deserves attention: a vintage model with expert knowledge and flexible opportunities for investment which will be helpful in examining issues related to technical progress.

Economic Policy Analysis

The models developed in this book are used to examine (i) technical progress, (ii) international trade and cooperation, and (iii) uncertainty and risk behaviour.

1. We show that and how, depending on the assumptions regarding technical progress, the optimal steady state of CO_2 concentration may rise or fall with increases in the steady

state level of progress. Notably, an improved substitute for fossil fuels always reduces the long-run level of atmospheric CO_2; while an improvement in fossil fuel productivity may increase or decrease the level of atmospheric CO_2.

2. We show what the solutions are for a model with neutral, constant, and ongoing technical progress, and in which way higher levels of technical progress lead to lower long-run optimal levels of atmospheric CO_2.

3. Several trade issues regarding CO_2 pollution will be considered. We examine whether a country or group of countries concerned about increasing atmospheric CO_2 and trading in energy goods should increase its exports of fossil fuels in a competitive world market, and should tax exports of fossil fuels in a non-competitive market place. On which specific world supply and demand conditions should decisions be made to subsidize or not to subsidize exports of fossil fuel substitutes.

4. We examine several cases of international cooperation in controlling CO_2 accumulation. The base case is complete cooperation between two regions in maximizing consumption with complete awareness of the CO_2 problem. This case is compared with a situation in which no cooperation takes place until a critical CO_2 level is reached.

We explore why in the non-cooperative situation the critical level is reached sooner, even though the region concerned about CO_2 always emits less carbon than in the base case.

As we have so far established, a paradoxical situation can occur in which initial emissions of the two regions are lower in the non-cooperative case than in the cooperative case.

5. Substantial applications apply to uncertainty. We show that the results from studies of the optimal use of a resource which is in limited supply can be applied to the CO_2 problem. This similarity is important because the results regarding the use of limited resources are extensive and powerful. Then, using numerical examples, we show that an inappropriate treatment of risk can lead to significantly higher than optimal estimates of the desirable level of fossil fuel use.

6.   Departures from the reference model involve multiple state models.   The key difference between this model and the reference model is the possibility of improving the economy by investment in knowledge and physical capital. This allows the description of more realistic long term behaviour. We show that both a stationary equilibrium in which no growth occurs in the economy and a dynamic equilibrium in which the economy grows at a constant rate are possible. We expect interesting results that concern the impact of changes in the values of key parameters on the equilibrium growth rate. Some of the results which we derive from the model and which we consider more closely are:

-    increases in the productivity of the energy sector increases the capital to knowledge ratio in equilibrium and the economy's rate of growth;

-    an increase in either the social discount rate or the consumption elasticity of utility causes the economy to grow more slowly, and

-    an increase in the depreciation rate of capital lowers the capital to knowledge ratio and the equilibrium growth rate.

7.   We explore the properties of another model, allowing flexible technical progress, in which prices influence the pattern of technical progress and non-neutral technical progress is possible. This model has several interesting features. First, it is a vintage model, the fossil energy input required by a piece of capital is determined by the year in which it is purchased and cannot be changed after the capital is in place. Second, research and development can be performed to improve the productivity of new capital or to reduce its energy requirements, the output required, the research budget, and cost of energy and capital are all exogenous. The objective is to minimize the cost of production.

8.   In evaluating large-scale models of energy, economy, and CO_2 interactions we explore how CO_2 feedback effects, forms of optimization, uncertainty, technical change and organizational forms explicitly treated in our models could be effectively used to greatly enhance the explanatory power of large-scale models.

Description of Chapters

The book is organized in nine chapters.

The first chapter as an introduction deals with the background of global environmental externalities and the major substantive issues involved in characterizing and identifying the CO_2 problem and trace gas accumulation. Chapter 2 presents the simplest, one state variable control model of an economy in which pollution of the prescribed kind occurs. We also analyze specific forms of this one state variable model which permits to trace the transition path to long-run performance. We also use the models to study major structural features such as technical progress, international cooperation and uncertainty. These features will be taken up separately each in later chapters.

In Chapter 3 it is shown that results from studies of the optimal use of natural resources being in limited supply can be naturally extended to the carbon dioxide problem.

Using numerical examples it is observed that an inappropriate treatment of structural uncertainty can lead to a significantly higher than optimal estimate of the desirable level of fossil fuel.

In Chapter 4 a more complex model of the economy with endogenous, neutral technical progress is analyzed. This is an EEE model with several state variables in which the major difference from earlier models, in Chapter 2, is the possibility of improving the economy by investment in knowledge and physical capital. This allows a more realistic description of long-run behaviour and adjustment.

Furthermore, in Chapter 4 we construct a model which is very flexible in the pattern of technical development and examine the reaction of research to energy price changes and limits on energy use. A vintage model of technical progress is introduced in which fossil energy input required by a piece of capital is determined by the year in which it is purchased and cannot be changed after the capital is in place.

Following the simple aggregative models of long-run growth and environmental constraints developed in Chapters 2 and 4, we focus on a discussion of large-scale models of energy, economy and CO_2 interactions in Chapter 5. Our primary conclusions are that CO_2 feedback effects and simple forms of optimization could be added to conventional EEE models without much difficulty. They would greatly enhance the usefulness of these models.

Chapter 6 takes us a significant step further. It considers a dynamic integrated model of the economy and climate change.

The greenhouse effect is considered as an externality generated by production and manufacturing activities. A two-sector general equilibrium model is constructed to study the characteristics of the time paths of the world economy and the global average temperatures under the competitive equilibrium.

Chapter 7 discusses and assesses several neo-classical growth models under global environmental constraints. It explores the interrelations between capital accumulation, global waste disposal and economic growth in emphasizing that environmental pollution reduces the production possibilities of the aggregate economy, but that global pollution might not retard economic growth.

Chapter 8, complementing Chapter 7, discusses a specific optimal economic growth model with a finite time horizon where fossil fuel resources constitute inputs to the aggregate production function bound by a critical cumulative CO_2 budget. The problem utilizes standard optimization procedures such as Pontryagin's maximum principle and dynamic programming. All of the approaches so far have adopted a social planning focus toward these issues.

Chapter 9 provides a treatment of uncertainty in view of optimal statistical decisions and stochastic dynamic programs. Using a model of optimal statistical decisions it is shown when it pays to 'act and learn' and when to 'learn and act'.

The value of information in reducing uncertainty can be shown to be sensitive to accuracy and likelihood of scientific research results.

The results are extended for the dynamic inter-temporal decision situation when the value of new information is an outcome of an optimal stochastic dynamic program.

REFERENCES

Chiarella, C., "Optimal Depletion of a Non-renewable Resource when Technological Progress is Endogenous", in Kemp, M. C. and Long, N. V. (eds), *Exhaustible Resources, Optimality and Trade*, North Holland, Amsterdam, 1980, pp. 81-93.

Conrad, J. M. and Clark, C. W., *Natural Resource Economics*, Cambridge University Press: Cambridge, 1987.

Deshmukh, S. D. and Pliska, S. R., "Optimal Consumption of Non-renewable Resources with Stochastic Discoveries and a Random Environment", *Review of Economic Studies*, 50, July 1983, pp. 543-554.

Deshmukh, S. D. and Pliska, S. R., "Optimal Consumption and Exploration of Non-renewable Resources with Uncertainty", *Econometrica* 48(1), 1990, pp. 177-220.

Edmonds, J. and Reilly, J., "Global Energy and CO_2 to the Year 2050", *Energy Journal* 4(3), 1983, pp. 21-47.

Fisher, A. C., *Resource and Environmental Economics*, Cambridge University Press: Cambridge, 1981.

Gilbert, R. J., "Optimal Depletion of an Uncertain Stock", *Review of Economic Studies* 46, 1979, pp. 47-57.

Manne, A. S., "ETA-MACRO: A Model of Energy-Economy Interactions" in Charles J Hitch (ed.), *Modelling Energy-Economy Interactions*, Resources for the Future, Washington D.C., 1977.

Manne, A. S. and R. Richels, Buying Greenhouse Insurance, MIT Press: Cambridge, Mass. 1992a.

National Research Council (NRC), Carbon Dioxide Assessment Committee. Nierenberg, W. A. (ed.) *Changing Climate*, National Academy Press: Washington D. C., 1983.

Nordhaus, W. D., "Thinking about Carbon Dioxide: Theoretical and Empirical Aspects of Optimal Control Strategies". Cowles Foundation Discussion Paper No. 565, Yale University, New Haven, Conn. October 1980.

Nordhaus, W. D., "Managing the Global Commons: The Economics of Climate Change", MIT Press: Cambridge, Mass. 1994.

Perry, A. M. et al, "Energy Supply and Demand Implications of CO_2, *Energy Journal*, 7(12), 1982, pp. 991-1004.

Rose, D. J., Miller, M. M. and Agnew, C., *Global Energy Futures and CO_2 Induced Climate Change*. Report MITEL 83-015, MIT Energy Laboratory, Cambridge,

Massachussetts, November 1983.

Yohe, C. W., "The Effects of Changes in Expected Near-term Fossil Fuel Prices on Long-term Energy and Carbon Dioxide Projections", *Resources and Energy*, 6, 1984, pp. 1-20.

CHAPTER 1

ISSUES OF GLOBAL ENVIRONMENTAL ECONOMICS

1.1 THE SCOPE OF THE PROBLEM

The focus of the first chapter is to address and analyse some specific and unifying issues at the interface of energy use and environmental management from a global perspective. These issues will be attacked at various instances throughout the book.

We start from the fact that carbon dioxide (CO_2) in the atmosphere affects the radiation balance of the earth, and that increasing (CO_2) concentrations are expected to cause a warmer climate.

Carbon dioxide is relatively transparent to energy as sunlight, but reflects or traps a large portion of this energy when radiated from the earth as heat. An important development in climatic studies is the identification of other sources these include: chlorofluoromethanes (major sources: spray cans and other commercial uses), nitrous oxide (major sources: jets, cars and unknown sources), and methane (major sources: anaerobic fermentation in rice fields, and natural gas leakages). These gases are often referred to as trace or greenhouse gases (GHG). Uncertainty about future increases in such gases significantly complicate the prediction of future temperatures. The prediction of future temperatures is further complicated by diverse factors. Significant feedback effects are expected to accompany any direct effects of CO_2 heat trapping. Warming may change cloud, snow, and ice cover and alter the earth's albedo or brightness. Consequently, the reflection and absorption of energy may increase or decrease. Also warming may cause the release of additional GHG which is trapped in frozen soils, accentuating the GHG problem. In addition, uncertain, major climatic disturbances are foreseen which are not connected with CO_2. Continental drift, fluctuations in the earth orbit, solar flux variations, and volcanic dust are all natural causes of climatic change. The release of particulates and changes in land-use are human activities which may impact local or regional climate.

We look into the suitability of energy resources, alone or in combinatiion, to satisfy these global environmental constraints, and identify reasonable scenarios for environmentally benign energy futures.

The last century has witnessed an unprecedented period of growth, in energy, economy, population and consumption. By the end of the sixties, in particular, anxiety was mounting about whether the world was beginning to collide with the ceiling of resource and environmental constraints, or whether it still had time to complete the gradual transition toward a "steady state" in which population and resources were in balance. The "limits to growth" debate culminated in several studies associated with the Club of Rome, predicting a catastrophic increase in mortality rates throughout the world beginning in the first decades of the 21st century, largely as a result of resource shortages. But even if resources were not limiting, global pollution would, a few years or decades later, lead to even greater catastrophe. The basic thesis was that the longer this catastrophe was postponed by "technological fixes" the more destructive would be the final collapse.

Consumption growth, as these studies argued, would so pollute the environment as to threaten human health and the life-supporting properties of the biosphere on which human existence depends.

There are three separate issues involved here:

(i)   general chemical or radiation contamination of the environment (hazardous and nuclear wastes)

(ii)   general deterioration of natural ecosystems (air and water pollution, soil erosion)

(iii)   changes in the global climate. (CO_2 emission and trace gases, ozone depletion).

Changes in the global climate, if they occur - offer now the most challenging task for the control of global environmental pollution. We particularly focus on (iii) as a problem of energy production.

Most studies so far have singled out carbon dioxide (CO_2) as the most serious pollutant and with potentially irreversible consequences on the global climate. (IPCC, 1990).

So far, no economically feasible control technologies to capture any significant fraction of CO_2 emissions have been on the horizon. A prime characteristic of the CO_2 problem is the long time lags that may elapse between the cause and the identification of significant effects. Another characteristic is the high degree of uncertainty that usually attends predictions of future effects. Such uncertainty is compounded by the fact that CO_2 effects occur at various complex levels.

The first level of effect is the direct physical result of the activity, in this case the actual CO_2 concentration in the atmosphere. The principal evidence for a trend toward increasing CO_2 levels in the atmosphere comes from continuous observations over many (up to thirty) years, at various sites.

The second level addresses the partitioning of CO_2 among the atmosphere and other carbon reservoirs. If we look at the carbon cycle, with the exception of the atmosphere, the amount of carbon in each reservoir is somewhat uncertain. It seems there is a natural circulation of carbon among these different reservoirs, particularly through photosynthesis and oxidation.

This opens a biogeochemical perspective on the carbon cycle. We can take the view that various chemical reservoirs of the earth are comparable to the organs in a human body. [Ausubel (1980)]. Accordingly, the CO_2 problem emanates from human activities feeding carbon into circulation faster than the ability of the earth's organs to digest or metabolize it. This is largely due to the increasing consumption of fossil fuels and less so to the clearing of natural forests. Carbon dioxide emissions from fossil fuels combustion have been growing at the rate of 4.3 percent per annum since 1860 (with the exception of the periods of world disaster, World Wars I and II and the depression years of the 1930s [Rotty (1977), Flohn (1989))].

We shall briefly describe the constraining factors that are the "givens" of our economic modelling. In Figure 1, curve A represents the present world energy use, unrestrained by CO_2 considerations. Curve B is a different world, where CO_2 concentrations satisfy two constraints:

(1) $M_{c1} \leq M \leq M_{cu}$

(2) $\dfrac{1}{M}\dfrac{M}{\delta t} \leq km,$ where m is a rate to be determined, and k being some fixed

parameter.

3

The first constraint to be determined on **M**, represents a ceiling level that might be approached with "acceptable" cumulative consequences, but which probably should not be exceeded. In addition, this level should be below any "threshold" level which heralds catastrophic events.

The second constraint is a rate limitation which depends on the rate of climate change, reflecting in a general fashion the assumption that slow rates of change are more amenable to compensating adjustments in human institutions and in productive activities that depend on climate.

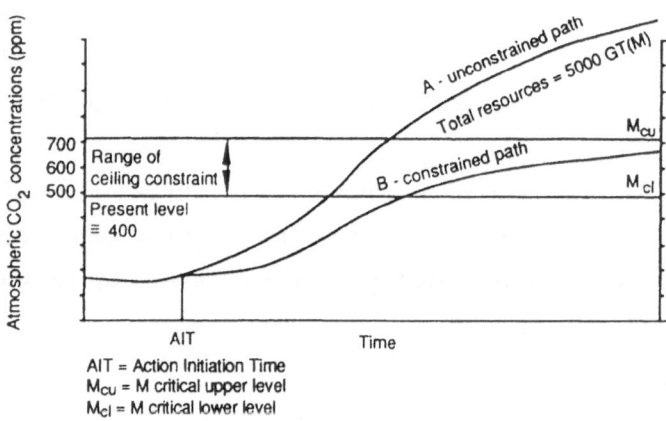

Figure 1: Schematic illustration of two CO_2 paths: „A" and „B"

1.2 WORLD FOSSIL FUEL USE

Given the CO_2 constraints described in the previous section we will analyze plausible world fossil fuel paths with the aid of a simple carbon cycle model. The model will be used to probe the sensitivity of the results to variations in key parameters that would influence the energy system dynamics. From the economic point of view of fossil fuel use the argument is often advanced that market forces should (and will) play a dominant role in allocating resources, and thus no other mechanism should intervene in this "optimum" process. According to this view, the "cost effectiveness" process is automatically operative, hence there is no reason for constraining exogenously the growth of fossil fuel use.

To explore these ideas further, note that the resource base for fossil fuels is indeed very large if it includes resources that will cost much more than they do now. About half of the conventional oil and natural gas is recovered inexpensively. The rest comes from a variety of sources that are associated with higher production costs, poorer fields that require drilling more holes, production from continental shelves, deeper basins and polar regions etc. All these activities are, moreover, associated with significantly larger environmental impacts. If we assume that cleaning up spills, reclaiming mined-out lands to useful purposes, and hazardous waste management are all operations whose cost must be internalized, then these "dirty" fuels become even more expensive.

The standard observation of resource economists e.g. that one never runs out of a resource but simply reaches a point where it is cheaper to use a substitute, is to some extent correct in the case of fossil fuels. Unfortunately, very few reliable estimates of fossil fuel resources versus cost curves are available.

Plans for increased energy use yield various scenarios for increased CO_2 concentrations with time. To capture in simple terms the nature of the problem we assume:

1. A minimum realistic level M_{cl} of CO_2 exists, determined by total integrated demand for fossil fuels added up over all future time, before the world has changed over to energy sources that do not increase CO_2 after that (principally solar and/or nuclear in various forms).

    We assume that if less total fossil fuels are used during all future time, there will be considerable social unrest due to unmet demand.

2. A maximum allowable level M_{cu} of CO_2 exists, determined by unacceptable climatic and/or biological and/or botanical effects. If it is exceeded, bad things happen.

5

3. We can plan our global fossil energy use so that the CO_2 Curve F increases smoothly to some value that is acceptable.

The curvature of the CO_2 buildup with time implies that society thought ahead, planned accordingly and avoided the trouble. Do we really have to plan ahead, even in the face of uncertainties, or can we let matters take their normal course, and later take crash measures when scientific research has finally shown with "certainty" that we should not exceed a given maximum level of CO_2? That is, can we behave so that the concentration increases, then stop in time?

It appears that we cannot do that because a discontinuous and sudden stop in the use of fossil fuel is practically impossible, it would imply a complete shift in technological infrastructure away from fossil fuel.

Technological breakthroughs notwithstanding, this is impossible for the world as a whole.

In order to describe the dynamics of carbon accumulation meeting the CO_2 ceiling constraints we assume the rate of buildup of CO_2 in the atmosphere to be represented by the following equations.

$$\dot{M}(t) = k\dot{F}_{fo}(t) + 1\,\dot{A}(t) - R(t)$$

where the dots denote time derivatives

$M(t)$ is the accumulated mass of carbon in the atmosphere (in tons) at time t,

$F_{fo}(t)$ is the rate of fossil fuel use (TWyr/yr)

k is the average emission rate of carbon per unit of fuel energy

$1\,\dot{A}(t)$ is the net release rate of carbon from clearing tropical forests

$\dot{R}(t)$ is the net transfer rate of carbon from the atmosphere to other reservoirs (sinks).

The objective is to find $F_{fo}(t)$ that conforms to a prescribed asymptotic M_α.

Unfortunately, most terms in the equation are only vaguely known.

l A(t) has been estimated, in the past few years with significant variations. The estimates are based on forest inventory statistics with fluxes calculated from a forest clearing rate times estimated carbon stored in particular forests.

R(t) represents the sinks of carbon dioxide. Currently, only the deep ocean is regarded as the ultimate sink of carbon.

Comparing fossil fuel use and the reconstructed and actual atmospheric record suggests an "airborne fraction" (AF) that according to most calculations with cycle models (Rosenberg et al., 1988), is between 0.55 and 0.65 depending on the oceanic vertical mixing rate and the behaviour of the biosphere.

One can establish an approximate relation between the atmospheric mass of carbon M(t) and the fossil fuel burning rate, $F_{fo}(t)$:

$$F_{fo}(t) \cong \frac{1}{K \cdot AF} \dot{M}(t) \tag{1}$$

This relation is the starting point for the development of the model.

We shall assume that the carbon in the atmosphere (added since Δt_o) can be modelled as a logistic function in the form:

$$M(t) = \frac{\beta}{1 + \gamma e^{-\alpha(t-t_0)}} + \lambda \tag{2}$$

where $\beta, \gamma, \lambda, \alpha$ = constants

t = time

t_o = action initiation time (AIT)

Since $M(t_o) = M_o$ it follows that:

$$\lambda \equiv M_0 - \frac{\beta}{1 + \gamma}$$

A ceiling constraint on CO_2 is imposed:

$$M_{cl} \leq M_c \leq M_{cu}$$

The model developed above attempts to describe a plausible world response to the impending CO_2 problem. The final world trajectory or response function could be thought of as a composite of three stages: during the first describing the period up to AIT, the response function follows closely the reference scenarios, hence it represents a period in which fossil fuel use is unconstrained by consideration of the CO_2 buildup. The AIT is however not the time when discussions or negotiations to limit CO_2 begin, but rather the time when the resulting policies begin to be effective. The second phase starting at the AIT signals the retardation of fossil fuel growth (because of the CO_2 build-up) and ends with fossil fuel use at its peak. The third phase of the response describes a "decay" and general "phase out" of the fossil fuel consumption.

1.3 TECHNOLOGICAL PATTERNS

Historical data on world energy consumption, when plotted versus time as a fractional share of different primary energy sources, follows a very regular pattern. This observation has given rise to the hypothesis that primary energy sources are competing for market shares, in producing intermediate and final products. Substitution of an existing technology or industrial processes to satisfy a given need by a new emerging technology has been the subject of a large number of theoretical and empirical studies.

One general finding is that almost all binary substitution processes (e.g. steam engine versus electric power), expressed in fractional terms, follow characteristic S-shaped curves which have been used for forecasting future competition between the two alternative technologies.

(This way to look at technological substitution processes has never been in much favour among economists because they feel that such processes lack a choice-theoretic or explanatory basis, see Stoneman, 1981.)

However, if we can find an endogenous explanation for the way technologies emerge and proliferate we will be able to parametrize such processes along the line of some logistic substitution model.

Most of the studies of technological substitution are based on the use of the logistic function. The logistic function, however, is not the only S-shaped function, but it is perhaps the most suitable one for empirical studies. Another S-shaped function, the Gompertz curve, has also been frequently used, especially to describe population, plant, and animal growth.

Following the work of Griliches (1957), Mansfield (1961) developed a model to explain the penetration of an innovation. He suggested that the penetration is directly proportional to the difference between the expected profit and expected investment associated with the innovation.

However, the first systematic attempt at forecasting technological change was made by Fisher and Pry (1970). This work extended Mansfield's findings, but considered only fractional shares of a market controlled by two competing technologies.

Let f be the total market share.

Most of the modelling approaches can be summarized by considering the following fundamental model

9

$$\frac{df}{dt} = (\beta + \alpha f)(f_m - f) \tag{3}$$

$$f(t=0) = f_o$$

where α, β = constants

$\qquad f_m$ = upper limit on market share

The constant α can be interpreted as the index of influenced adoption so that $\alpha f(f_m-f)$ can be thought of as the imitation component. Similarly, β can be interpreted as the index of uninfluenced adoption, so that $\beta(f_m-f)$ can be explained as the innovation component.

The Fisher-Pry model can be obtained from (3) by setting $\beta = 0.0$ and $f_m = 1.0$; that is

$$\frac{df}{dt} = \alpha f(1 - f) \tag{4}$$

The differential equation has the solution:

$$\ln\left(\frac{f}{1-f}\right) = \alpha t + \gamma, \quad \text{or} \tag{5}$$

$$f = \frac{1}{1 + \exp(-\alpha t - \gamma)} \tag{6}$$

This process may be thought of as a pure imitation diffusion process.

In (4), if $f = 0$, so is $\frac{df}{dt}$. Thus its validity requires that the process has risen above some initial threshold by other means. This implies that $f(t_0) > 0$. That is, in order for diffusion to occur according to the model, there must be a finite proportion of the population that has already adopted the innovation.

(3) with $\beta = 0$ can be solved for f(t):

$$f(t) = \frac{f_m}{1 + \left(\frac{f_m - f_0}{f_0}\right) e^{-\alpha f_m(t - t_0)}} \tag{7}$$

where $f_0 = f(t_0)$

To find the maximum diffusion rate, set $df/dt = 0$, which gives for t*,

$$t^* = t_0 - \frac{\ln f_0 - \ln(f_m - f_0)}{\alpha f_m}$$

hence $f(t^*) = f_m/2$. Also evaluating the change in t*, as α varies yields,

$$\frac{dt^*}{d\alpha} = \frac{lnf_0 - \ln(f_m - f_0)}{\alpha^2} \tag{8}$$

Since $0 \le f_0 \le 1$ it follows that $\frac{dt^*}{d\alpha} < 0$ so that any increase in the imitation index α, will increase f(t) for all t > t_0; that is, it will shift the logistic curve to the left.

(3) is also a first order differential equation. Its unique solution can be found by specifying an initial condition,

$$f(t) = \frac{f_m - [\frac{\beta(f_m - f_0)}{\beta + \alpha f_0}] e^{-(\beta + \alpha f_m)(t - t_0)}}{1 + \left[\frac{\alpha(f_m - f_0)}{\beta + \alpha f_0}\right] e^{-(\beta + \alpha f_m)(t - t_0)}} \tag{9}$$

Note that f(t) is not symmetrical about t*, since f(t*) = $f_m/2$ - β /2α, the majority of the adoptions in the diffusion process given by (3) will occur after the rate of diffusion achieves a maximum.

Starting with (5), t_m can be calculated as follows:

$$t_m = [\ln(\frac{0.5}{1 - 0.5}) - \ln(\frac{0.01}{1 - 0.01})]\alpha^{-1}$$

From the preceding discussion it is prudent to assume that new technologies that will be necessary as future complements to and replacements for fossil fuel will not be introduced faster than technologies were introduced in the past. Also note that the dynamics of technological substitution is more complex than depicted here. Since the penetration rate tends to decrease with decreasing energy growth rate it can be assumed that future penetration rates will be slowed correspondingly. In particular, we need to explain penetration rates endogenously, a part of an interaction of economic factors within an economic model. This will be pursued later (Chapter 4).

11

1.4   UNCERTAINTY ANALYSIS

The high levels of uncertainty have prompted some policymakers and scientists to adopt the "wait and see" approach, since no agreement or even a process towards "optimal" strategy is on the horizon. Thus, they caution against any major quick actions until better data and information is available (Chapter 7). It is, however, quite possible that none of these substantial uncertainties will be reduced in the next decade (Ausubel, 1980).

In other words, the conventional assumptions of learning over time may be irrelevant in this issue. It is possible that we will face virtually the same decision in a decade that we face now, with only slightly more reliable information. This is so partly because the climate change is plagued with the problem of "indivisibilities" and "scant sets" (Olson, 1982). These are areas where our knowledge is generally so meager and the stakes are so high, but also that we cannot expect to get reliable answers either cheaply or quickly because a decisive experiment entails a policy change and, because historical experience with scant sets is so slowly informative. It is clear that decision-makers must attempt to live with and accommodate to the high degree of uncertainty permeating the CO_2 issue. Therefore, policymakers must realize that decisions will have to be made without certain cost/benefit information and that waiting until the uncertainty is resolved will be probably too late to make an effective decision.

More attention should be directed at resolving or reducing the effects of "malign uncertainties" (uncertainties which make it impossible to determine whether the outcome of a particular policy will be reasonably good or very bad), as opposed to resolving "benign uncertainties" (those which make it impossible to tell which of two good strategies will be better) (Burgess, 1980). This means that uncertainties and therefore losses become asymmetric around the optimal decision and have to be treated accordingly (Morgan and Henrion, 1990).

1.5 DESIGN OF AN INTERNATIONAL REGIME

Undoubtedly, for the CO_2 issue and many other "global commons", some form of international regime is not only desirable but necessary. Design of operative mechanisms for a CO_2 regime can be partially extracted from studying other mechanisms developed at the regional or international level to deal with "similar" situations.

The envisioned international CO_2 regime requires much more substantial global agreement, control, management, and allocation of resources. Most of the international agreements discussed have failed to reach substantial agreement (beyond agreeing to study the matter further, monitor, and so forth). Even if substantive agreements were reached, there is real grounds for doubting that paper agreements are being seriously enforced.

The relative lack of enforcement machinery internationally necessarily conditions the character of international negotiations. Thus, it is often less costly and easier to hold meetings, conferences and other forums and even draft "statements of principles" than to actually do something substantial. We already are experiencing this in the context of the proliferation of meetings and papers on the CO_2 issue.

The difficulties arising from investing a CO_2 international regime with requisite responsibilities and authority and even legitimate command of coercion stems directly from the reluctance of nations to yield sovereignty to international bodies, for it means some loss of control over decisions that directly affect important national interests and domestic constituencies. However, in practice, some delegation of responsibility cannot be avoided and a variety of means are attempted by each nation to make it more acceptable.

A serious investigation of "optimal" strategies should be initiated under which we can organize reasonably effective coordinated efforts among sovereign states to reach substantive agreements on the CO_2 issue. Policies should attempt to find powerful self-interest and incentives which could be a mixture of selfish and social motivation, to foster an acceptable agreement. The fact that diverse opinions and high uncertainties surrounding the relative redistribution of the climate on the regional level might be helpful for achieving early consensus on action.

The actual tactical steps required to initiate an early international regime for CO_2 should be explored very carefully with an eye to the requirements of political legitimacy and the necessity for efficiency and internal consistency in decision-making on highly complex issues.

The above conclusions lead to the following policy-oriented recommendations:

1.   International cooperation is needed in many areas in order to ensure timely and orderly "transition" to inevitable non-fossil energy systems. Areas of cooperation include:

(a)   More efficient and rational worldwide energy use and perhaps subsidies of certain end-use-efficient technologies.

(b)   Arresting the increasing fuelwood crisis in the LDCs.

(c)   International R&D in non-fossil energy systems, development, and commercialization. Attention should be given to avoid "monism" in energy policy. All feasible options should be pursued.

(d)   Establishing a world energy financing centre with low interest loans to aid the LDCs in technology acquisition, adaptation and training.

2.   It should be of the highest priority to incorporate the CO_2 issue into the world energy policy-making process and CO_2 should serve as the discriminating factor in the choice of energy options.

3.   A serious investigation to establish an international institution to deal with the CO_2 issue should be promptly initiated. Consideration should be directed to the following:

(a)   How best to optimize effective coordinated efforts among sovereign states.

(b)   Setting of standards or "acceptable" CO_2 limits that will serve as policy targets.

(c)   A system of carbon "quotas" or allocation schemes.

(d)   Coordinated scientific cooperation in data gathering, monitoring and emission regulation.

(e)   International agreement on enforcement machinery.

(f)   The needs of the LDCs, their ability to substitute, and their developmental objectives.

4.   The US, EC, and the OECD countries should plan in the next decade to take early action at limiting their fossil fuel use. This is required regardless of the CO_2 problem.

5.   There is a need to develop guidelines for world coal trade and reach early agreements toward "internationalization" of the trade, especially with the signs of the second coming of the "coal age".

In concluding, it is appropriate to recognize the limitations of the analysis; partly due to the paucity of relevant data, the constellation of uncertainties pervading all aspects of the CO_2 problem, the lack of an integrative framework, and above all the "social engineering" ethical, value-laden implications of the issues raised.

1.6   ADAPTATION

Adaptation is perhaps the path of least resistance with respect to expected changes in climate (Schelling, 1992). Qualitatively, it can easily be shown that migration and industrialization will be the basic implication of any climate change - especially a slow one - as it impacts upon agriculture. Change in agricultural productivity means that the ratio of population density per unit of agricultural productivity is changed, and such changes can be compensated by corresponding changes in the population density (migration) or by increasing agriculture or other economic activities (reeducation and industrialization).

Explicit government intervention to change behaviour could also assist in the adaptation. For example, zoning laws could be planned to restrict new buildings from being constructed in areas of probable flooding because of projected sea level rise. However, recent efforts to prevent construction in flood plains have been notably unsuccessful. Historically, people do not behave the way "foreseen" by advocates of adaptation. A good example of how myopic the adjustments might turn out to be was presented by Ronald Ridker (1981). Ridker described how coastal populations are likely to adjust to a slow but inevitable rise in sea level. In view of the slowness of the change (on the order of a few feet each century), and man's tendency to use a positive discount rate, it is more likely that sea walls and dikes will be built than that people will evacuate. And once such sea walls are built, it will appear cheaper to make them a bit thicker and higher than to evacuate the area. Eventually, much of the human race could find itself living below sea level, with the probability of a catastrophic breach in the dikes growing over the centuries. This is an example of a situation in which man's normal response to adaptation could eventually become self-destructive.

There are considerable costs involved in this adaptation process as well. It might include major new investment, significant change in production methods, development of new business relations, and disruption and dislocation of present human settlements. Adaptation is inherently redistributive, and hence could be inequitable especially for the developing countries. Most of the developing countries, lacking strong technological infrastructures, are highly vulnerable especially to changes in agriculture and water supply. Thus, adaptation could very well accentuate the North-South cleavage. Adaptation seems to be working well and is well founded by economic theory (see Chapter 2) when climate change is only moderate, but is much more difficult to justify if changes are significant because of second-order more severe consequences (Nordhaus, 1991; Cline, 1992; Fankhauser, 1994).

REFERENCES

Ausubel, J., "CO_2: An Introduction and Possible Board Game". Working Paper WP-80-153, International Institute for Applied Systems Analysis (IIASA), Laxenburg, Austria, 1980.

Burgess, C. and H. Burgess, *Energy Decision Making Under Conditions of Uncertainty*. MIT Energy Lab., Working Paper No. MIT-EL80-042 WP, Cambridge, 1980.

Cline, W., *The Economics of Global Warming*, Institute of International Economics, Washington D.C., 1992.

Fankhauser, S., Valuing Climate Change. The Economics of the Greenhouse Effect, Earthscan: London 1994.

Fisher, J. C. and R. H. Pry, "A Simple Substitution Model of Technological Change." Report 70-C-215, General Electric Co., Research & Development Center, Schenectady, N.Y., Technical Information Series, June, 1970; see also, *Technological Forecasting and Social Change 3*, 75 (1971).

Flohn, H., "Treibhauseffekt der Atmosphäre: Neue Fakten und Perspektiven" ("Greenhouse Effect of the Atmosphere: New Facts and Perspectives"), Rheinisch-Westfälische Akademie der Wissenschaften, Vortrage N 379, Sept. 27, 1989.

Griliches, Z., "Hybrid Corn: An Exploration in the Economics of Technical Change," *Econometrica 25*, 501-522 (1957)

Climate Change, The IPCC Scientific Assessment, Cambridge University Press: Cambridge, 1990.

Mansfield, E. "Technical Change and the Rate of Innovation", *Econometrica* 27(4). 1061.

Morgan, M. G. and M. Henrion, *Uncertainty: A Guide to Dealing with Uncertainty in Quantitative Risk and Policy Analysis*. Cambridge University Press, Cambridge, 1990.

Nordhaus, W. D., "Economic Policy in the Face of Global Warming", in J. W. Tester et al. (eds), *Energy and the Environment in the 21st Century*, M.I.T. Press: Cambridge, Mass. 1991, 103-118.

Olson, M., "A Conceptual Framework for Research about the Likelihood of a Greenhouse Effect", in: US Department of Energy, Carbon Dioxide Program, Washington DC, 1982.

Ridker, R. G. "Social Responses to the CO_2 Problem," in USDOE Carbon Dioxide Effects: Research and Assessment Program, *Proceedings of the Carbon Dioxide and Climate Research Program Conference*, Washington, D.C., 1981, p. 227.

Rotty, R. M., "Global Carbon Dioxide Production from Fossil Fuels and Cement, AD 1950 - AD 2000", in *The Fate of Fossil Fuel CO_2 in the Oceans*, Anderson et al., eds., Plenum Press, New York 1977, 167-181.

Schelling, T., "Some Economics of Global Warming", *The American Economic Review 82(1), 1992, 1-14.*

Stoneman, R., *The Economic Analysis of Technology*, Oxford Univ. Press: Oxford 1981.

CHAPTER 2

ECONOMIC MODELS OF OPTIMAL
ENERGY USE UNDER ENVIRONMENTAL CONSTRAINTS

2.1   INTRODUCTION

The major purpose of this chapter is to contribute to better policy making through improvements in models studying the economic impacts of the carbon dioxide problem, and to show ways in which economic instruments can effectively be put to use to alleviate such a problem.

This approach differs in at least one major aspect from common studies of the climatic change problem. We focus on the analysis, control and optimization of modelling forms rather than the collection and analysis of data. More concretely, we search for optimal fossil fuel use, research and technology policies rather than predicting the future. Most studies of the problem exogenously specify technical developments and fossil fuel control policies and then predict future climatic changes. These prediction models incorporate a great deal of data and tend to be quite complex.

The advantage of an optimizing control model is increased flexibility in structural and dynamic assumptions on the economy allowing explicit 'what-if' questions to be asked about the possibility of controlling the growth in atmospheric CO_2 concentrations.

Let us start by looking at a class of single state aggregate optimal control models. They allow consideration of static production but also technical change. In the latter case we take care of the fact that very small rates of ongoing technical change can have an enormous impact because of the very long time-span associated with the CO_2 problem, that is the 100-150 years until major effects occur.

In this class of single state models, the level of atmospheric CO_2 is the only state variable. The only policy or control variable chosen is fossil fuel use. (In follow-up chapters 4 and 5 we will explore a class of multiple state models including additional state variables, stocks of physical capital and levels of knowledge).

The simplest type of technical change is a finite or limited improvement in a technology. Because such a change is not ongoing, the model remains static and relatively easy to examine. Ongoing but uncontrolled technical change is also examined with a single state model.

Some major policy conclusions can be derived from this class of model:

1.   One can show that depending on the assumptions regarding technical progress, the optimal steady state CO_2 concentration may rise or fall with increases in the steady

state level of progress. Notably, an improved substitute for fossil fuels always reduces the long-run level of atmospheric CO_2, while an improvement in fossil fuel productivity may increase or decrease the level of atmospheric CO_2.

2. Solutions of a model with neutral, constant and ongoing technical progress and the basic static model are very similar. In the model with technical progress, higher levels of technical progress lead to lower long-run optimal levels of atmospheric CO_2.

3. We examine two cases of international co-operation in controlling CO_2 accumulation. The base case is complete co-operation between two regions in maximizing consumption with complete awareness of the CO_2 problem. This case is compared with a situation in which no co-operation takes place until a critical CO_2 level is reached. Our most important finding is that in the non-co-operative situation the critical level is reached sooner, even though the region concerned about CO_2 always emits less carbon than in the base case.

4. The last applications are on uncertainty. We first show that the results from studies of the optimal use of a resource which is in limited supply can be applied to the CO_2 problem. This similarity is important because the results regarding the use of limited resources are extensive and powerful. Then, using numerical examples, we show that an inappropriate treatment of uncertainty can lead to significantly higher than optimal estimates of the desirable level of fossil fuel use.

2.2   SOME ECONOMIC STUDIES ON THE CO_2 PROBLEM

Two major kinds of economic studies can be identified for dealing with the CO_2 problem. The first category treats economic and economic modelling issues in the context of an integrated framework of energy-economy and climate changes, the second category applies the theory of resource use and depletion to the management of CO_2 emissions.

In what follows we provide a selected overview of those studies which yield pertinent results comparable to our own or which are of methodological interest in modelling the energy-economy-climate interactions, a more extensive review is presented in Chapter 5.

A few years ago the US National Research Council (1983) compiled a detailed investigation that up to now constitutes the most extensive, comprehensive and consistent examination of the climate change problem. It uses energy-economy, climate and agricultural models to predict future impacts of carbon dioxide and trace gas accumulation. The major conclusions are that no radical actions should be taken, that increases in carbon dioxide are likely, and that more research is necessary. In this report, the developers of the energy-economy model (W. Nordhaus and C. Yohe) note that the technology development and elasticity of substitution parameters critically affect the model's results.

It should be added that the method of modelling technical change and energy substitution possibilities is also critical and controversial. The economic modelling chapters of the NRC report have been updated by Yohe (1984). The conclusions of the report have not changed significantly.

Another collaborative study, the joint MIT-Stanford study (Rose, Miller, and Agnew, 1983), is of interest because it is one of few reports which search for alternatives to increasing CO_2 and offer some positive choices. In the Edmonds and Reilly model (Edmonds and Reilly, 1983) S-shaped paths are exogenously specified for several new energy technologies. The MIT-Stanford study modifies these paths and also looks at additional technologies. It finds that the adoption of realistic CO_2 reducing technologies, while not eliminating a significant CO_2 warming could increase the CO_2 doubling time to several centuries.

In attempting to discuss optimizing strategies W. Nordhaus (1980) made a seminal contribution by applying simple optimization models to the qualitative and quantitative analysis of the CO_2 problem. By letting the consumption equation depend on fossil energy use he determines the appropriate tax policy to control CO_2 and makes a quantitative estimate of this tax.

In most present studies of the CO_2 problem we find that technological change and technology substitution are specified exogenously or modelled in a very simple fashion.

In this regard, the logistic (S-curve) assumption on the diffusion of new technologies has become very popular, although it lacks sufficient economic explanatory power. For example, in a study by Perry et al (1982) an energy demand level and a fossil fuel use pattern is assumed. Fossil fuel use follows a logistic curve between the present and an assumed ultimate level of use. The rate of non-fossil energy growth needed to fill the gap, between fossil energy use and assumed total energy demand is then examined. This study emphasizes the importance of analysing the investment needed in non fossil energy to fill this gap but it does not present a model of the substitution process. In the model we suggest, the substitution process is a direct result of our maximization of welfare.

Modelling the impacts of changes in energy use on the economy is a major problem. A good starting point for such considerations would be the ETA-Macro model (Manne, 1977) though it has not been used for studying the CO_2 problem. This model can be described as a multisector, forward-looking model. It examines consumption and investment policies and their impact on national welfare. National welfare is measured by discounting utility from the present to a distant horizon. ETA-Macro consists of two models: a macro-model of the whole economy and a more detailed model of the energy sector. The model seems more sophisticated in its treatment of capital and the determination of the desirable level of energy use than those economic models presently used in CO_2 analysis. However, it is limited to the USA in geographic scope which makes it unsuitable for examining international problems such as CO_2. The model has no endogenous technical progress which we consider an important feature for the analysis of CO_2. On the other hand, the model has several features which would be desirable in models of energy, economy and the environment. It is optimizing and considers costs and benefits of capital investment.

Very recently, there has been a flurry of economic modelling and assessment studies of CO_2 effects, in particular by Manne and Richels (1990), Peck and Teisberg (1991) and Nordhaus (1993), (1994).

In setting up this approach we were influenced by applications of optimal control models to pollution problems (Fisher, 1981; Conrad and Clark, 1987). Such models often show different structures, e.g. pollution affecting utility or production, the pollutant acting as a stock or a flow and the way abatement activities are available.

In this specific context, we find that many models and results on the depletion of a non-renewable resource could be applied with simple modifications to the CO_2 problem. Under two assumptions the problem of fossil fuel use in the face of increasing carbon dioxide is parallel to the problem of consumption of a limited resource. The first assumption is that the carbon dioxide absorption rate is sufficiently small to be ignored. The second is that CO_2 impacts follow a "step" pattern: that is, CO_2 has no impact on productivity until a critical level, M_c, is reached. Then if the CO_2 level exceeds M_c production falls to zero, or remains stagnant. One of the most serious effects in facing global warming is that of irreversibility, that is, given the accumulation of atmospheric CO_2 we will reach a critical level of the CO_2 budget where there is a point of no return (unless technologies are in place that effectively remove CO_2 from the budget). The interpretation of a critical level of atmospheric CO_2 accumulation where there is a precipitous drop of production means that we have reached the biophysical limits of growth.

An interesting treatment of endogenous neutral technical progress in a depletion model was suggested by Chiarella (1980). He proves the existence of a steady state growth path and a simple rule governing the rate of investment in research. Research investment along the optimal path should be carried out until the growth rate in the marginal accumulation of technology equals the difference between the marginal product due to an extra unit of research investment and the marginal product of capital.

A similar problem is the use of a limited non-renewable resource when the reserve of the resource is unknown. The model by Gilbert (1979) can be directly converted to a model of fossil fuel use when the critical CO_2 level is uncertain. Under the above assumptions this problem is equivalent to determining the rate of fossil fuel use when the critical concentration of atmospheric carbon dioxide is unknown. The results show that the optimal use of fossil fuel is lower when uncertainty is properly considered than when the expected values are assumed to be certain.

Deshmukh and Pliska (1980, 1983) study more complex models of the same problem. The possibility of doing exploration to find new reserves is a significant addition in their models. The parallel in the CO_2 problem is research to increase the probability of finding a technology for the removal of CO_2 from the atmosphere. Their findings imply that in the periods between discoveries or research breakthroughs, fossil fuel use and consumption fall, but if research is very successful long-run fuel use may rise.

2.3 PRELIMINARY DEFINITIONS AND THE GENERAL MODEL

Our interest in model structure and policy options for energy policies leads us to the use of forward-looking, optimizing aggregate models. The class of models we are analysing here uses mathematical techniques of optimal control (Kamien and Schwartz, 1981, 1991). This requires: a measure of welfare or benefit as an objective function, and a definition of how policies, the control variables, affect specific aspects of the world, the state variables. A complete optimal control model of the CO_2 problem is an ideal rather than a reality, although outlining a detailed model provides a reference when examining simple models focusing on specific issues.

The basic ingredient of our model is an uncomplicated measure of welfare denoted by J. J is the sum over all time of the discounted flow of welfare. U is the utility or the flow of welfare at any instant. The social discount rate is r. Utility depends only on consumption, C. Consumption, production, and investment all have the same, single measure. Production is designated Y. There are numerous investment possibilities represented by the vector $\underline{I}$. At least two world regions are imagined and trade can occur between them. The vector of traded goods is $\underline{X}$, and the prices of the goods are in the vector p. Consumption equals production minus the sum of investments in capital goods plus the returns from trade. Production depends on current inputs, capital inputs, atmospheric CO_2, and the use of imports. Current imports are not stored and affect production as a flow. Capital inputs can be accumulated and affect production as a stock, $\underline{K}$. We distinguish two types of current inputs, fossil fuels, $\underline{F}$, and all other current inputs, $\underline{E}$. Atmospheric CO_2 is a single number denoted by M. The welfare measure, J, can be expressed as:

$$\text{Max} \quad J = \int_{O}^{\infty} e^{-rt} U(C(t)) \, dt \qquad (1)$$
$$\underline{F}, \underline{E}, \underline{I}, \underline{X}$$

Production and consumption are determined by

$$C = Y - \underline{1} \underline{I} + \underline{p} \underline{X} \qquad (2)$$

$$Y = f(\underline{F}, M, \underline{E}, \underline{K}, \underline{X}) \qquad (3)$$

26

and $\underline{1}$ is the unit vector.

We assume that one region has the leadership in world trade, and its decisions on trade influence world prices through a vector function

$$\underline{p} = g(\underline{X}) \tag{4}$$

The regional use of fossil fuels, $\underline{F}_i$ also reacts to the trade policy, this reaction is determined by the vector function:

$$\underline{F}_i = h(\underline{X}) \tag{5}$$

The capital stocks and the level of atmospheric CO_2 are the state variables. Knowledge is treated as a special type of capital stock which does not depreciate. The current capital stocks and the CO_2 level may affect their own rate of change through depreciation and reabsorption, respectively.

The dynamics of their change are expressed as follows:

$$[d\underline{K}/dt] \equiv \dot{\underline{K}} = j(\underline{I}, \underline{K}) \tag{6}$$

$$[dM/dt] \equiv \dot{\underline{M}} = k(\underline{F}, \underline{F}_i, M) \tag{7}$$

The general model introduces many of the concepts to be elaborated: the importance of a welfare measure, feedback to production from increasing CO_2, changes in physical capital and knowledge, the ability to control these changes, and the impact of trade on fossil fuel and production. In its general form the model is too complex to solve for relevant results; therefore, we develop a set of simple models to examine these concepts. The simplification is achieved in two ways: by reducing the number of variables and restricting the functional forms.

27

2.4 A SIMPLIFIED MODEL

This model contains only the most important elements of the general optimal control model introduced in the previous section. As in the general model, the economy considered has only one consumption good, C, and a flow of utility, U(C), results from consuming this good. The objective is to maximize, J, which equals the utility flow discounted at the rate r and integrated from the present time, $t=0$, to infinity.

We assume that production of the good, C, depends only on the use of fossil fuels, F, and the level of carbon dioxide accumulated in the atmosphere, M. Control of fossil fuel use is the sole means of managing the economy. No decisions on investment and trade, as in the general model, are made. Thus, the assumptions are:

$$J^* = \underset{F}{Max} J = \underset{F}{Max} \int_{0}^{\infty} e^{-rt} U(C) \, dt \tag{1}$$

$$C = f(F, M) \tag{2}$$

We assume that production is finite at any finite level of fossil fuel use and that the production function is continuous in F.

The equation for change in the atmospheric carbon dioxide level is more specific than in the general model. We assume that each unit of fossil fuel use emits a fixed amount of CO_2 into the atmosphere; therefore, fossil fuel use and CO_2 accumulation can be measured in similar units. Finally, CO_2 leaves the atmosphere naturally at a rate proportional to the CO_2 concentration. The proportionality factor or "reabsorption rate" is α. We specify the relations which determine M:

$$dM/dt = F - \alpha M \tag{3}$$

$$M(0) = M_0 \tag{4}$$

This model already contains three specific (economic) assumptions. First, neither emissions nor carbon dioxide accumulation impact utility (comfort or health) directly; the impacts occur through productivity changes. Second, the accumulated CO_2 is the pollutant not the rate of CO_2 emissions, differentiating CO_2 from such pollutants as SO_2 or particulates. Third we

28

assume a very simple equation for CO_2 accumulation. Both the retention of CO_2 in the atmosphere and its reabsorption may in fact be nonlinear and change over time. The model differs from the earlier general model in two fundamental ways: there are no opportunities for investment and there is only one world region.

Several additional assumptions pertain to utility and production. Utility increases with consumption, but at a decreasing rate:

$$U' > 0 \quad \text{and} \quad U'' < 0 \tag{5}$$

where primes stand for derivatives.

Whatever the level of CO_2 accumulation, we assume that production increases with increases of fossil fuel use (close to zero) and that the rate of increase in production slows with additional fossil fuel inputs. But production declines with the accumulation of CO_2. Thus, the conditions hold with fossil energy being burnt.

$$\partial f(0, M) / \partial F > 0 \tag{6}$$

$$\partial^2 f / \partial F^2 \leq 0 \tag{7}$$

$$\partial f / \partial M \leq 0 \tag{8}$$

$$F \geq 0 \tag{9}$$

2.4.1 Necessary Conditions

First establish the Hamiltonian H:

$$H(F, M, \phi) = U(C) + \phi(F - \alpha M) \tag{10}$$

The Hamiltonian equals the utility flow plus the current value of increasing carbon dioxide concentrations. The adjoint value ϕ represents the marginal value of increasing CO_2 concentrations. The term λF is added to form the Lagrangian L, and assures that $F \geq$

29

0 :

$$L = H + \lambda F \tag{11}$$

From (11) we derive the necessary conditions

$$\partial L/\partial F = U' (\partial f/\partial F) + \phi + \lambda = 0 \tag{12}$$

$$d\phi/dt = r\phi - \partial L/\partial M = (r + \alpha) \phi - U' (\partial f/\partial M) \tag{13}$$

$$\lambda F = 0, \ \lambda \geq 0 \tag{14}$$

Since (12) implies that ϕ is less than zero, this agrees with our intuition that the value of increasing CO_2 should be negative. In order to deal with a CO_2 induced cost, we define q equal to $-\phi$, and q can be referred to as the shadow price of CO_2 emissions. We can now restate (12) and (13) as

$$U' (\partial f/\partial F) - q + \lambda = 0 \tag{15}$$

$$dq/dt = (r + \alpha) q + U' (\partial f/\partial M) \tag{16}$$

or

$$q = \int_{t}^{\infty} -e^{-(r+\alpha)(\tau - t)} U' (\partial f/\partial M) \, d\tau$$

(15) and (16) are crucial conditions that lend themselves immediately to economic interpretations. (15) implies that fossil fuels are used up to the point at which the marginal contribution to utility equals the shadow price, unless fuel use is forced to zero. Unless the marginal utility is increasing, shadow price increases will drive down the use of fossil fuels. The condition in (16) is easily understood in its integral form. Increases in atmospheric CO_2 lower productivity and thus cause a disutility. The cost of CO_2 is the discounted sum of the marginal harm or disutility due to an increase in CO_2. We discount at the rate α because a unit of CO_2 emitted presently disappears at the rate of α, and we discount at r to put future

30

losses on a present value basis.

A further necessary condition determines maxima or minima.

By defining $U_{ff} = U'(\partial^2 f / \partial F^2) + U''(\partial f / \partial F)^2$ (17)

the necessary condition for a maximum can be stated as

$$\partial^2 H / \partial F^2 = U_{ff} \leq 0 \tag{18}$$

2.4.2 Sufficient Conditions

We assume that the first and second derivatives of the production function are defined. (In problems where the derivatives are not defined sufficiency must be proved by other methods)[1]. We use sufficiency conditions that require that the necessary conditions are met and that the utility function is jointly concave in M and F. The proof only applies when the state equation (3) is linear in F and M.

The definition in (17) and the two following makes the statement of concavity more concise:

$$U_{mm} = U'(\partial^2 f / \partial M^2) + U''(\partial f / \partial M)^2 \tag{19}$$

$$U_{fm} = U'(\partial^2 f / \partial F \partial M) + U''(\partial f / \partial F)(\partial f / \partial M) \tag{20}$$

The concavity conditions can be stated as follows:

$$U_{ff} \leq 0 \tag{21}$$

and

$$U_{ff} U_{mm} - (U_{fm})^2 \geq 0 \tag{22}$$

From (22) to assume concavity of the utility function the second partial derivative of f with respect to M must be less than or equal to zero, this assumes that U_{mm} is less than or equal to zero. This requirement matches the assumption of many scientists that CO_2 impacts will

31

accelerate at higher levels of CO_2. Another sufficiency condition is simply that q(t) does not get too big. This condition can be stated as:

$$e^{-rt} q(t) \to 0 \quad \text{as } t \to \infty \tag{23}$$

The condition is satisfied if q has a finite equilibrium value or if q grows at a rate less than r. The conditions in (15), (16), (21), (22) and (23) are sufficient to assume that an optimal path for the control variable has been found.

2.4.3 Definition and Optimality of Equilibrium

In the context of controlling fossil fuel use, the notion of equilibrium is interesting because it predicts the distant future, indicates the general direction of movement from the present to the long run.

The equilibrium is defined in terms of q and M. To specify the equilibrium conditions we assume that for F greater than zero that F can be found (as a function of q and M from (15) and (16)). This function is specified as $\psi(q, M)$. F is constant when q and M are constant. From (3) if M is constant

$$\psi(q, M) - \alpha M = 0 \tag{24}$$

Equation (24) defines combinations of shadow price, q, and CO_2 concentration M which keep the concentration of CO_2 constant. M is greater than zero, therefore, from (24), F is greater than zero in equilibrium.

Setting dq/dt equal to zero over time in (16) and substituting for F gives a second condition on the equilibrium

$$(r+\alpha) q + U'(\partial f(\psi(q, M), M)/\partial M) = 0 \tag{25}$$

(25) constitutes something like a price/damage equation. It assures that in equilibrium the higher the incremental damage of CO_2 the higher the price of fossil fuels.

If the equilibrium satisfies (15), (16), (21), (22) and (23) it is optimal. q is finite in

32

the equilibrium satisfying the sufficiency condition in (23). If we assume that the production function has a curvature there satisfying (21) and (22) we know the equilibrium is optimal.

Equilibrium condition (25) can be restated in terms of the more familiar rate of substitution

$$-(\partial f/\partial F)/(\partial f/\partial M) = dM/dF = 1/(r+\alpha) \tag{25'}$$

Along an isoquant dM/dF, the slope of the isoquant equals the negative of the ratio in (25'). In Figure 1, the curve ab is the locus of all points such that $dM/dF = 1/(r + \alpha)$ along an isoquant. These points represent an efficient balance between the marginal gains from increased emissions and losses from increased atmospheric CO_2. $1/(r + \alpha)$ can be thought of as the price ratio of the value of increased fossil fuel use to decreased CO_2 levels. OC is the line along which F equals α M, the set of stationary points. The equilibrium is at the intersection of curves ab and OC in the figure. The line OC always has a steeper slope than dM/dF because r is greater than zero. Higher values of r lower dM/dF and move the equilibrium along OC. Because of the relative slopes of ab and OC, this means we move to higher levels of atmospheric CO_2 and lower levels of long-run production as r increases. Not surprisingly, a high discount rate causes us to value long-run consumption less.

2.4.4 Illustration by a Phase Plane Diagram

For conventional production functions, the equilibrium can be represented in a phase plane diagram as in Fig 1a. The phase plane diagram is a valuable tool not only because it shows the equilibrium but because it shows the changes in variables over time. In Appendix A, we precisely define the stability and existence conditions which assure that Fig 1a represents the equilibrium, that a path to the equilibrium exists, that the path is unique, and that therefore an optimal equilibrium is a long-run optimum.

Equations (24) and (25) are plotted in Fig 1a as curves AB and CD respectively. On the left-hand side of Fig 1a the curve AB lies above CD; the curves slope toward each other and intersect at equilibrium, 0. The relation between the curves results from our previous assumptions regarding the slopes of the production function, f(F, M). If the shadow price is above AB, fossil fuels are relatively expensive, fossil fuel use is reduced, and the CO_2 level

33

increases for the opposite reasons.

These movements are indicated by the small arrows in the diagram. To maintain the steady state, that is to satisfy (24), fossil fuel use must be low when the CO_2 level, M, is low and high when M is high.

Therefore, along AB when M is near zero the steady state shadow price, q, is large lowering fossil fuel use, and, at higher values of M, q is lower. Curve CD traces the "price vs harm" equation, (25). Along this line the shadow price of CO_2 emissions equals the long-run harm due to a marginal increase in CO_2 concentration. To the left of this curve, dq/dt is positive and to the right dq/dt is negative, again illustrated by the arrows. If the CO_2 concentration is low, M to the left of CD, the harm due to CO_2 is low relative to the future impacts; therefore, the shadow price of emissions is increasing. The opposite effects occur to the right of CD. With low historic levels of atmospheric CO_2, CO_2 increases have caused little harm. This implies that, at low levels of M, the marginal harm due to CO_2 is low; and to satisfy (25), the shadow price, q, must also be low. As M increases along the "price vs harm" curve, CD, q also increases. The curved arrows in the phase plane diagram describe the change in variables over time. When the equilibrium meets the sufficiency conditions noted earlier and the phase plane can be illustrated as in Fig 1a, it is optimal to choose the unique q so that the level of atmospheric CO_2 increases monotonically towards the equilibrium from levels of CO_2 less than the equilibrium. FO in Fig 1a represents such an optimal path. This means that both the shadow price, q, and atmospheric CO_2, M, increase with time. The use of fossil fuels, F, decreases monotonically, as can be seen by taking the total differential of (15).

34

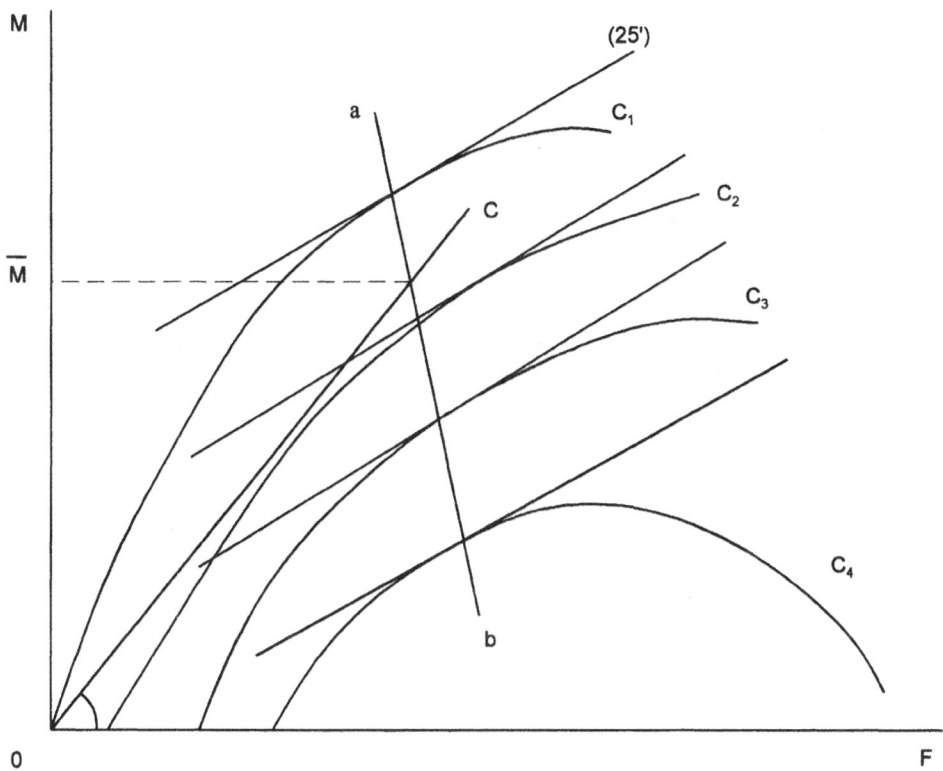

Figure 1: Solution of Model

35

$$AB: \psi(q,M) - \alpha M = 0 \qquad CD: -(r+\alpha)q + U' \frac{\delta f(\psi(q,M),M)}{\delta M} = 0$$

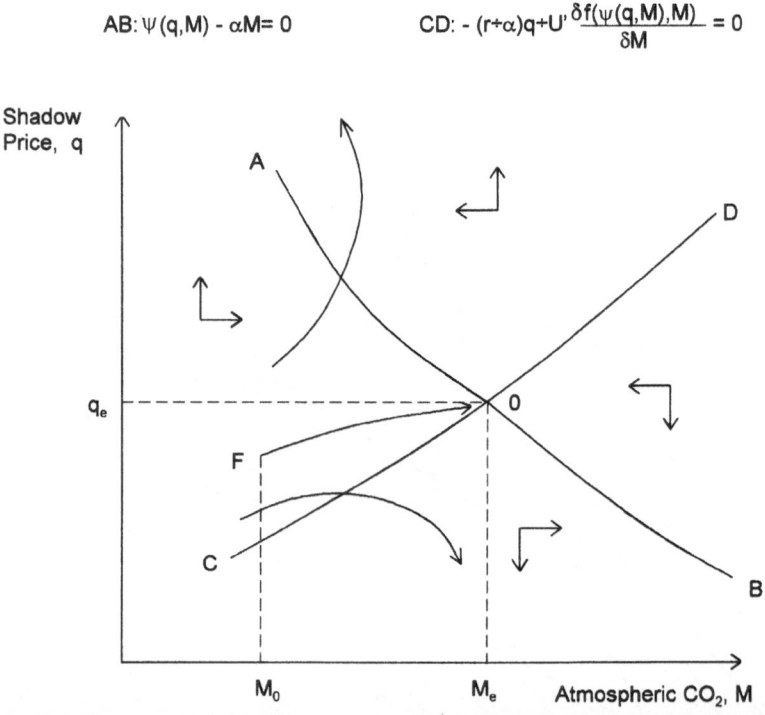

Figure 1a: Phase Plane Diagram

2.5 A DISCRETE TYPE IMPACT OF CO₂ EMISSIONS

In this section we analyse, as a more specific form of the model, the negative impact of CO_2 emissions that occur abruptly at specific levels of atmospheric CO_2.

In the simplest way we describe how consumption C is affected by CO_2 accumulation:

$$C = \begin{cases} f(F), & M \le M_c \\ 0, & M > M_c \end{cases} \tag{1}$$

f being a production function with $f'(0) > 0, f'' < 0$ which means that, at zero fossil fuel use, fossil fuels are always productive, although they may be unproductive at higher levels of use. The basic assumption is that carbon dioxide has no impact on production until it reaches a critical level M_c. First assume that if CO_2 levels exceed M_c, production falls to zero and the step function's simplicity has very drastic consequences. Damages from CO_2 accumulation could rise rapidly at a critical level.

The necessary conditions for the basic problem when C is given by (1) are:

$$q = \begin{cases} U'f', & M \le M_c \\ 0, & M_c < M \end{cases} \tag{2}$$

$$\frac{dq}{dt} = (r + \alpha) q, \quad M \ne M_c \tag{3}$$

Because the optimization is over an infinite horizon, there is no simple necessary transversality condition on the adjoining variable, q.

The problem is further complicated because the derivative of the Hamiltonian with respect to M does not exist at M_c. However, a careful analysis of the problem can determine q and F at M_c. If the function f gives a maximum at F_m and F_m is less than or equal to αM_c, there is a simple answer. Producing at the maximum for all time creates the highest possible utility. If producing at the maximum never raises M above M_c, we set q equal to zero and F at F_m. Both (2) and (3) are then satisfied. If F at the maximum of f is greater than αM_c or if f has no maximum, the solution is more complex. But it can be shown that it is always optimal to use the entire CO_2 capacity, that is burn fossil fuels in a manner which raises atmospheric CO_2 to the critical level M_c.

The analysis first determines that F equals αM_c when M equals M_c. Second, q is determined by (2), (3) and the additional necessary condition that q(t) is a continuous function (except where M is constrained).

The calculation of q is simplified by finding T in the time horizon at which M equals M_c. The existence, uniqueness and finiteness of T is then proved which directly leads to the conclusion that the optimal solution exists and is unique.

The following heuristic considerations lead to this result. If M equals M_c, F must be less than or equal to αM_c; if not, M becomes greater than M_c and production drops to zero. The properties of U and f assure that (2) can be inverted and a function g, with g(q) equal to F exists, which is monotonically decreasing in q. (2) further assures that q is greater than or equal to zero. (2) and (3) together assure that dF/dt is less than or equal to zero for M less than or equal to M_c. If F is strictly less than αM_c, then αM_c, or equivalently, q is strictly greater than $U'f'(\alpha M_c)$, F would remain less than αM_c for all times. Such a path is dominated by many alternative paths including F equals αM_c. Therefore, F equals αM_c when M equals M_c. This is pretty much in the spirit of Krelle's (1987) description of an ecological equilibrium though obtained from different model reasoning. These observations can be summarized in a proposition whose statement and proof is left for the Appendix B.

T is unique and exists. Assuming an optimum exists, it follows that the q_o, F and M which satisfy the necessary conditions are all unique, exist and are optimal. Figure 2 illustrates typical paths of fossil fuel use and CO_2 accumulation in the step model. M rises and F falls over time, both reach their equilibrium values at T.

The assumption that production falls to zero when the critical CO_2 level is reached is very extreme. We describe a model of slightly greater complexity that avoids this extreme assumption:

$$C = \begin{cases} f(F), & M \leq M_c \\ \beta f(F), & M > M_c \end{cases} \tag{4}$$

where $\beta \leq 1$.

In the previous model we did not require that f have a maximum. However, in this model we must assume a maximum of f exists or the integral in the Appendix B equation (B3) is unbounded. We assume that F_m greater than αM_c maximizes f.

38

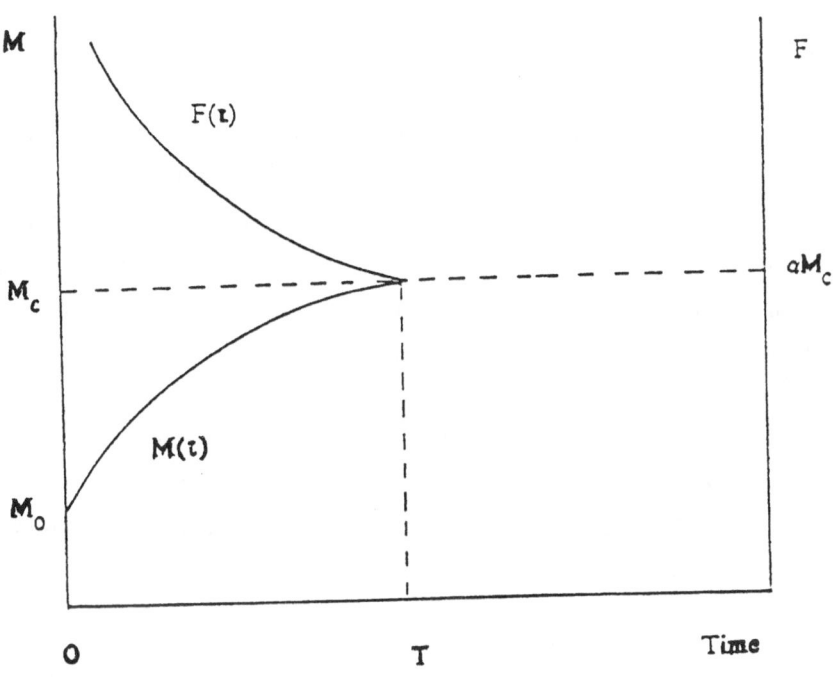

Figure 2: Illustration of Paths of F and M in Step Model

39

The necessary conditions are

$$q = \begin{cases} U'f', M \leq M_c \\ \beta U'f', M > M_c \end{cases}$$ (5)

$$\frac{dq}{dt} = (r+\alpha) q, \quad M \neq Mc$$ (6)

In addition q(t) must be continuous.

Again the discontinuity of production at M_c complicates the problem. However, by examining paths of q which satisfy the three necessary conditions the possible solutions can be reduced to two.

If an optimum exists (5) assures that q is greater than or equal to zero; further, q is less than or equal to $U'f'(\alpha M_c)$ as shown in the previous case. If the value of q at time zero, q_0, is zero, then (6) and continuity assure that q equals zero for all t. F equals F_m for all t, M equals M_c at some finite time, and M goes to F_m/α as t goes to infinity. If q is greater than zero, q rises exponentially until q equals $U'f'(\alpha M_c)$. q rising further is non-optimal; therefore, M must equal M_c when q equals $U'f'(\alpha M_c)$. The problem is identical to that presented in the previous case.

Consider the optimal action when M_0 equals M_c. F equals either αM_c or F_m for all time; therefore utility flow will be constant for all time at either $U(f(\alpha M_c))$ or $U(\beta f(F_m))$. The value of F which maximizes output will be optimal. If $f(\alpha M_0)$ is greater than $\beta f(F_m)$, F equals αM_c is optimal. If $f(\alpha M_c)$ equals $\beta f(F_m)$, the actions are equivalent. If $f(\alpha M_c)$ is less than $\beta f(F_m)$, F equals F_m is optimal.

Figure 3 illustrates both the production function and the optimal solution. We consider two step sizes β_1 and β_2 . As illustrated by the dotted line (Curve A) and along the left axis, $\beta_1 f(F_m)$ is greater than $f(\alpha M_c)$; therefore it is optimal to use fossil fuels at F_m. As illustrated by the dashed line (Curve B and along the left axis), $\beta_2 f(F_m)$ is less than $f(\alpha M_c)$, therefore, in this second case long-run fossil fuel use is αM_c.

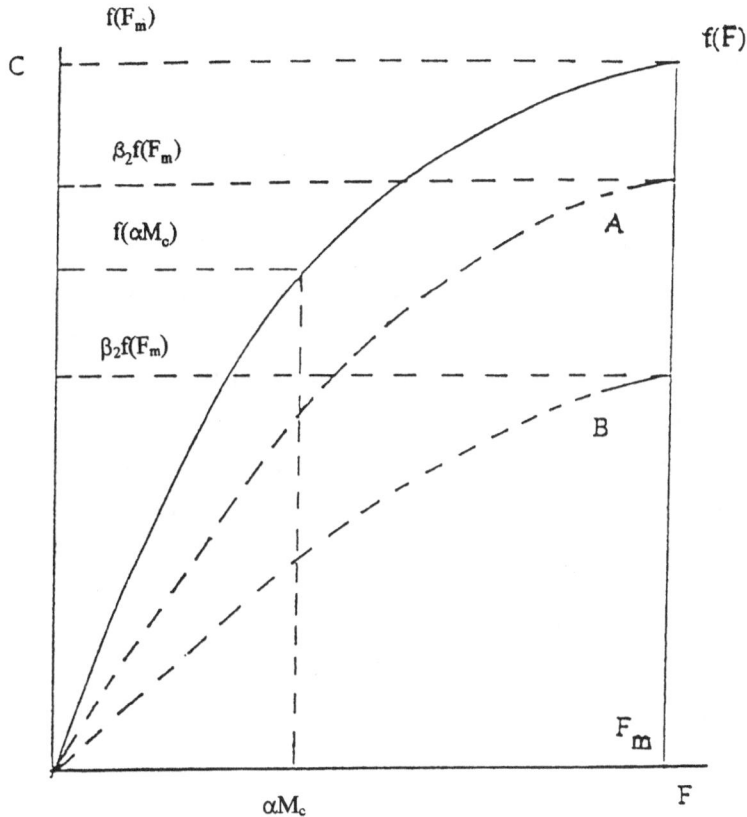

Figure 3: Two Levels of CO₂ Impact

2.6 FURTHER SPECIFICATION OF THE MODEL

Further specification of the model allows the sensitivity analysis of particular variables. Two major results emerge: an approximation of F as a function of M and a demonstration of the importance of the productivity and efficiency in the fossil fuel sector in determining the present optimal fuel use. We specify U(C) and f(F) by:

$$U(C) = \begin{cases} \dfrac{C^{1-\gamma}}{1-\gamma}, & \gamma \neq 1 \\ lnC, & \gamma = 1 \end{cases} \tag{1}$$

$$f(F) = \lambda F^{\epsilon} \tag{2}$$

These functional forms have all the properties formerly assumed for U(C) and f(F), lead to a simple solution, and are familiar in the economics literature. The elasticity of consumption and production with respect to fossil fuel use (i.e. the per cent change in consumption and production over the per cent change in fossil fuel use) is constant at ϵ, referred to as the fossil fuel productivity factor.

The utility function is sometimes referred to as the constant relative risk aversion utility function. γ is the consumption elasticity of utility or the relative risk aversion. Equation 5 (2) takes the form:

$$\left. \begin{array}{ll} M \leq M_c, & \epsilon \lambda F^{-(1-\epsilon)(1-\gamma)} \\ M > M_c, & 0 \end{array} \right\} = q \tag{3}$$

Then using the other necessary conditions, F(t), as a function of αM_c is found:

$$F(t) = \alpha M_c e^{A(T-t)} = F(0)e^{-At} \tag{4}$$

where

$$A = \frac{r + \alpha}{1 - \epsilon(1-\gamma)} \tag{5}$$

Note that F(t) is independent of the scaling factor λ .

Equation (4) for F(t) can be substituted into equation (B1) (Appendix B) and a simple

42

expression in T derived:

$$\frac{\alpha M_c}{(A-\alpha)M_o}(e^{(A-\alpha)T}-1)+e^{-\alpha T}(M_c/M_o)=0 \tag{6}$$

A comparative statics analysis of (6) shows that: T decreases when the social discount rate or the fossil fuel productivity, r and ϵ respectively, increase: T increases when the consumption elasticity of utility, γ increases; and the change in T with changes in the rate of CO_2 absorption α, is indefinite.

The key equation is the simple approximation for F, the emissions policy, derived from (4).

$$F=A(M_c-M)+\alpha M \tag{7}$$

This shows that unless A is much greater than α, the present policy is sensitive to all the model's parameters. A comparative statics analysis of (7) shows how the fossil fuel use pattern reacts to parameter changes within the range of interest. Highlights of the main impacts can be summed up:

- Because of the lower value of future consumption when the discount rate is high, initial fossil fuel use will increase with increases in the social discount rate.

- A higher consumption elasticity of demand, γ , tends to reduce present fossil fuel use. In general, a higher γ moves the economy to a more uniform consumption pattern. In this case, since consumption is falling, a higher consumption elasticity of demand reduces present fossil fuel use and increases future fossil fuel use.

- An increase in reabsorption, α, increases present fossil fuel use. When reabsorption is high, present emissions of CO_2 have less impact on future atmospheric CO_2 levels, encouraging higher present fossil fuel use.

- The initial use of fossil fuels increases with increases in the fossil fuel productivity factor, ϵ. This can be best understood by relating the fossil fuel productivity factor to the price elasticity of demand for the fossil fuel input which we call σ . σ

43

equals $1/(1-\epsilon)$, thus in this model high productivity implies high price elasticity of demand. From 5(6) the shadow price of fuels increases at a rate dependent only on the social discount rate and CO_2 reabsorption; therefore, the higher the price elasticity the more rapidly the fossil fuel use declines. A high rate of decline implies high initial use and a rapid approach to the equilibrium.

Values for T and $F(0)/M_c$ for a variety of parameter values are given in Table 1. The table shows the very significant impact which the fossil fuel productivity factor has on the optimal present level of fossil fuel use and the time to equilibrium. A high productivity factor results in very high fossil fuel use and a very short time to equilibrium. This suggests both that if the productivity is high it is not optimal to stringently restrict present fossil fuel use and that if fossil fuel productivity is low, it is important to conserve fossil fuel and delay the time at which very low levels of fossil fuel use are necessary.

Table 1: T and F(0) for a Variety of Parameter Levels

$\epsilon(1-\gamma)$ $\epsilon(1-\gamma)$	α	r	M_c/M_o	T	$F(0)/M_c$
0.5	0.001	0.02	2	77.44	0.0205
0.95	0.001	0.02	2	13.80	0.1915
0.5	0.002	0.02	2	69.87	0.0210
0.95	0.002	0.02	2	12.48	0.1830
0.5	0.001	0.06	2	34.48	0.0605
0.95	0.001	0.06	2	5.40	0.0592
0.5	0.002	0.06	2	28.96	0.0610
0.95	0.002	0.06	2	4.88	0.5830
0.5	0.001	0.02	8	91.25	0.0351
0.95	0.001	0.02	8	15.26	0.3344
0.5	0.002	0.02	8	75.51	0.3525
0.95	0.002	0.02	8	14.01	0.3188
0.5	0.001	0.06	8	39.12	0.1051
0.95	0.001	0.06	8	5.88	1.0344
0.5	0.002	0.06	8	33.59	0.1053
0.95	0.002	0.06	8	5.36	1.0188

2.7 AN EXTENDED COBB-DOUGLAS MODEL

In the discrete type model the impacts of CO_2 manifest themselves suddenly, in contrast, an extended Cobb-Douglas type model shows the negative effects occurring gradually. We use the definition of utility presented in 6(1). Again, the model assumes that a critical level of CO_2 accumulation, M_c, exists at which production drops to zero.

M_c is used to define two variables, F_s and M_s: $M_s = M_c - M$ and $F_s = F - \alpha M_c$.

M_s gives the distance to the critical level of CO_2 concentration, and F_s is the distance from the level of emissions which maintains the critical CO_2 level. Consumption increases with increases in either of these variables. The production/consumption function is:

$$C = F_s^\epsilon \, M_s^{1-\epsilon} \tag{1}$$

We refer to this model as an extended Cobb-Douglas model because equation (1) is a Cobb-Douglas form. From (1) it is clear that it will never be optimal to exceed the critical level of atmospheric CO_2 and that the CO_2 level will always be increasing because F will be greater than αM_c.

An assumption inherent in (1) is that the economy produces less and less over time.

M_c and ϵ are the significant parameters of the model. For F much greater than αM_c, ϵ is approximately the fossil fuel elasticity of production, however, as F approaches αM_c the elasticity of production, however, as F approaches αM_c the elasticity of production with respect to fossil fuel use approaches infinity. The model is limited in that the elasticity of production with respect to CO_2 concentration is completely determined by ϵ. Using the variables M_s and F_s, the equation that describes the dynamic changes in M can be restated as:

$$d M_s / dt = - (F_s + \alpha M_s) \tag{2}$$

The necessary conditions for the problem are:

$$\epsilon C^{-\beta} \, [M_s / F_s]^{1-\epsilon} = q \tag{3}$$

with ß as the consumption elasticity of utility.

$$\frac{dq}{dt} = (r + \alpha) q - (1-\epsilon) \, C^{-\beta} \, [F_s/M_s]^\epsilon \tag{4}$$

45

In the model, F_s and M_s equal zero at the equilibrium or steady state. At the equilibrium, there is no production or consumption. Marginal utility is infinite while total utility is minus infinity (for $\beta < 1$). If the initial CO_2 level is less than the critical level, it is not optimal to reach the equilibrium. Fortunately, because of the familiar form chosen, we can guess a solution for the transition path. We assume that M_s changes at an exponential rate, g: Therefore,

$$M_s = M_s(0) \, e^{gt} \tag{5}$$

$$F_s = F_s(0) \, e^{gt} = -(g + \alpha) \, M_s(0) \, e^{gt} \tag{6}$$

$$C = C(0) \, e^{gt} = (-(g + \alpha))^\epsilon \, M_s(0) \, e^{gt} \tag{7}$$

From (6) and (7) it is obvious that this solution can only be optimal if g plus α is less than zero.

Examining (3) we see that q changes at a rate of $-\beta$ g. Using (3), (4) can be rewritten as

$$\frac{dq}{dt} = [r + \alpha \, \frac{(1 - e)}{e} \, \frac{F_s}{M_s}] q \tag{8}$$

By dividing q through (8), the left hand side becomes the rate of change in q, and the following expression is derived

$$-\beta g = r + \alpha \, \frac{(1 - e)}{e} \, \frac{F_s}{M_s} \tag{9}$$

By substituting for F_s/M_s and rearranging terms, we derive

$$-g = \frac{re + \alpha}{1 - e(1 - \beta)} \tag{10}$$

Because the parameters are positive the numerator is always positive in (10), and because ϵ and $1 - \beta$ are less than one the denominator is always positive. g is therefore always negative. This assures that consumption is always decreasing and that the integral of discounted consumption is limited. If g satisfies (10) the assumed solution is optimal. The similarity between the equation for g and the equation for A, 6(5), is obvious. If we rewrite (6) as

46

$$F_s(t) = F_s(0) eg^t \qquad (11)$$

the similarity between the path of emissions in this model and that in the discrete-type model (expressed in 6(4)) is evident. The fossil fuel policy can also be expressed in a manner parallel to equation 6(7).

$$F = -g(M_c - M) + \alpha M \qquad (12)$$

In this model, unlike the step model, the equilibrium is not reached in finite time. The restrictions of g and F to changes in the parameters are the same as the restrictions of A and F in the previous model. Increases in r, ϵ and α all increase the present level of fossil fuel use. An increase in β decreases the present use of fossil fuels.

2.8 DISCUSSION AND PERSPECTIVE

Simple step and Cobb-Douglas type models of CO_2 - economy interactions have been analyzed. Consumption equals F^ϵ in the step model and $A F_s^\epsilon M_s^{1-\epsilon}$ in the Cobb-Douglas model. The parameter A is for scaling so that the outputs of both models are equal at M equals one and F equals 0.01. For the Cobb-Douglas model, the consumption fossil fuel use relation changes at different CO_2 levels; therefore, consumption is plotted for M equal to one, two and three (Figure 4).

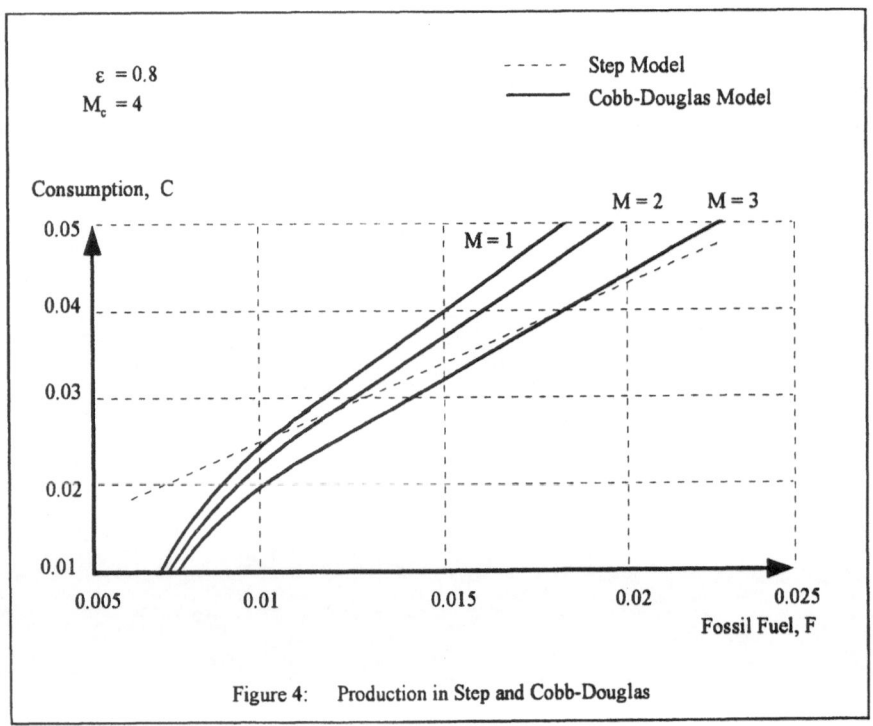

Figure 4:   Production in Step and Cobb-Douglas

In both models there is a critical carbon dioxide level at which consumption and production drop to zero, and this critical level is the equilibrium level of CO_2 accumulation. Furthermore, the optimal fossil fuel use path is a declining exponential. The level of present fossil fuel use responds to parameters similarly in the models: present fossil fuel use increases with increases in the discount rate, the rate of CO_2 absorption, and the elasticity of production with respect to the fossil fuel measure. The fossil fuel use reaches equilibrium in a finite time in a step model, but never reaches equilibrium in a Cobb-Douglas model (Figure 5).

The relation of current emissions to the critical carbon dioxide level is examined in both the step and Cobb-Douglas models. The parameters used in calculating the graphs are listed. Because the impacts of CO_2 in the Cobb-Douglas model occur more quickly, the initial emissions are lower in the Cobb-Douglas model.

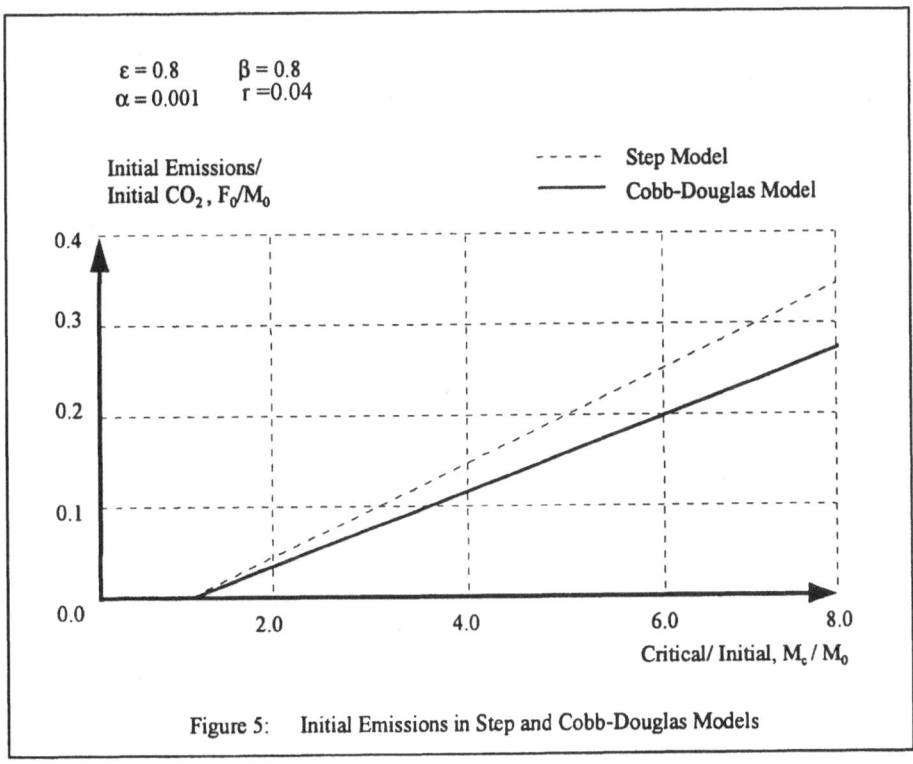

Figure 5:   Initial Emissions in Step and Cobb-Douglas Models

Small and simple models of this kind are useful for generating ideas, supplementing large-scale models, and improving modelling approaches.

Several areas of economic research will be particularly important to CO_2. One important area is the effectiveness of research support in changing the mix of energy technologies. Another area of research is the use of market controls such as excise taxes, export limits, subsidies and cartels; to reduce or increase the use of a commodity.

The CO_2 problem should be considered in determining the type of research in the energy area which government supports. From a CO_2 perspective, economic incentives for the development of coal and oil shale are questionable. It appears that improvements in all non-fossil fuels do not necessarily lower future levels of CO_2. To displace fossil fuels, alternative non-fossil fuels must be highly substitutable. This study suggests two other factors which policy makers should be aware of when considering the economics of CO_2. As with many other environmental problems, any policies which increase the perceived discount rate will exacerbate the CO_2 problem by reducing our concern about the future. The shadow price attributed to fossil fuels by expected value economic models will be lower than the true shadow price for a risk averse society.

In the remainder of this chapter we pursue policy-directed applications of the simple models proposed so far for a range of three major issues: technical progress, international co-operation, and uncertainty. Each of these issues will be treated in more detail in separate chapters.

Because of technical progress affecting energy production or finding mitigating strategies against CO_2 emissions (climatic engineering) the CO_2 problem may be highly sensitive to changes of technical progress parameters.

The worldwide nature of climate change, the dispersion of major fossil fuel resources, and the variation in possible effects of climate change all suggest the importance and problems of international co-operation in developing a CO_2 control policy.

Enormous uncertainty surrounds the future levels and effects of atmospheric CO_2 on various levels of model construction and analysis. A specific type of uncertainty, structural uncertainty, associated with the modelling process has impacts on present optimal policies which have not been previously considered.

2.9   A TAXONOMY OF TECHNICAL CHANGE

The long-run level of atmospheric CO_2 depends on both the degree of technical change and its form. In the following cases, technical change varies in its impacts on the productivity of inputs and the reabsorption of CO_2. We show that these differences create very different incentives for the long-run use of fossil fuels.

In an ongoing system of technical progress, the system of state equations has an additional parameter S, e.g. knowledge. Let a function of three variables represent consumption.

$$C = f(F, M, S) \qquad\qquad (1)$$

In this system of equations, the existence of equilibrium, its optimality, and stability are governed by the same conditions as in the previously presented simple models.

We examine several kinds of technical change. In all cases the equilibrium is examined under the previously stated conditions.

1.   Neutral technical progress

$$f(F, M, S) = S\, h(F, M) \qquad\qquad (2)$$

We refer to this as a neutral technical progress because the ratio of this partial derivative of f with respect to F and the partial derivative of f with respect to M is independent of the level of technical progress.

Examining condition 4(25') the constancy of this ratio under different levels of neutral technical progress assures that the long-run carbon dioxide concentration is invariant.

2. Development of non-fossil, substitutable energy

Along this line the production function represents technical progress in the development of a completely substitutable non-fossil fuel:

$$f(F, M, S) = h(F + S, M) \tag{3}$$

The equilibrium condition is the same condition as in 4(25') except that it is evaluated at F + S and M:

$$-[\partial h(F+S, M)/\partial F] / [\partial h(F+S, M)/\partial M] = dM/dF = 1/(r+\alpha) \tag{4}$$

dM/dF is defined along an isoquant of h.

The immediate impact of the level of technical progress can be found by comparative statics or by a graphic examination of the solution. By neglecting details it shows that more technical progress reduces the equilibrium level of CO_2 and fossil fuel use.

3. Removal of atmospheric CO_2

Removal of CO_2 changes the dynamics of CO_2 accumulation. Let S be the rate of CO_2 removal, the equation for the change in atmospheric CO_2 is:

$$dM/dt = F - (\alpha M + S) \tag{5}$$

The equilibrium condition is then $F = \alpha M + S$, and this case parallels the previous case with a similar interpretation.

4. Fossil fuel enhancing technical progress

Changes in fossil fuel productivity would result in a future production function:

52

$$f(F, M, S) = h(SF, M) \tag{6}$$

In equilibrium the following equation must hold along an isoquant of h

$$-S[\partial h(SF,M)/\partial F]/[\partial h(SF,M)/\partial M] = S(dM/dF) = 1/(r+\alpha) \tag{7}$$

Because the slopes of the isoquants may increase or decrease at any given value of F, the locus of points satisfying (7) may move to the right or left. The change in the equilibrium level of atmospheric CO_2 is, therefore, uncertain.

5. Emissions Purification

Scrubbing CO_2 from emissions changes the dynamics of CO_2 accumulation as

$$dM/dt = F/S - \alpha M \tag{8}$$

The equilibrium condition is then $F = S\alpha M$. This case corresponds to Case 4, and the same conclusion holds. The equivalence can be seen if the effects of substituting a variable F_o equal to F/S into the production function and (8) are considered.

6. Amelioration of CO_2 impacts (generated by forms of climatic engineering)

One representation of relieving CO_2 impacts would change the production function as

$$f(F, M, S) = h(F, M - S) \tag{9}$$

This case is similar to 2. Consequently, the long-run equilibrium level of CO_2 is raised with higher levels of technical progress.

A more general taxonomy is possible for production functions that are invariant under some general transformations to be called "neutral technical progress" (Sato, 1981). Such a taxonomy could translate itself into specific technology induced policies like reforestation, recovery of CO_2 from power plants, storage in the oceans,

53

disposal in depleted gas reservoirs, energy technology substitution etc (Okken et al, 1989).

In the context of a narrower focus, e.g. energy technology substitution, very recent studies by Manne and Richels (1990/1992a), on the basis of their Global 2100 model, have given rise to assessment of a broad spectrum of CO_2 benign technologies.

2.10 IMPACTS OF TECHNOLOGICAL CHANGE IN CO_2 EMISSION CONTROL

It has been shown that the long-run level of atmospheric CO_2 depends on both the degree of technical change and its form. A taxonomy of technologies has been given that vary in their impacts on productivity of inputs and the reabsorption of CO_2. Here we present a diagrammatic analysis on the impacts of technological change in CO_2 emission control.

As a classification we look at the following major effects of finite technical progress:

1. non-fossil substitute: $f(F,M,S) = h(F+S,M)$ (Sec. 9.2)
2. fossil fuel enhancement: $f(F,M,S) = h(FS,M)$ (Sec. 9.4)
3. impact amelioration: $f(F,M,S) = h(F,M-S)$ (Sec. 9.6)
4. exogenous neutral technical progress: $f(F,M,T) = h(F,M)S(t)$ (Sec. 9.1)

Case 1: Non-fossil, substitutable energy:

The development of a completely substitutable non-fossil fuel can be characterized by

$$f(F,M,S) = h(F+S,M)$$

The equilibrium condition of the basic reference model is evaluated at $F+S$ and M:

$$-\frac{\frac{\partial h(F+S,M)}{\partial F}}{\frac{\partial h(F+S,M)}{\partial M}} = \frac{dM}{dF} = \frac{1}{(r+\alpha)} \tag{1}$$

dM/dF is defined along an isoquant of h. In this formulation the impact of the level of technical progress can be found by comparative statics or by a graphic examination of the solution. Figure 6 shows the isoquants for the technology $h(X,M)$ where $X = F+S$. This type of technical progress changes the value of X at any point F along the horizontal axis, it essentially moves the vertical axis to the right at higher levels of productivity. In Figure 6, for technical progress equal to S_1, O_1C_1 is the stationary CO_2 line, $M/F = 1/\alpha$. At the higher level of technical progress S_2,O_2C_2 is the constant CO_2 line. Figure 6 shows that O_2C_2 cuts the balanced input line, ab, at a lower level, and thus that more technical progress reduces the equilibrium level of CO_2 and fossil fuel use.

55

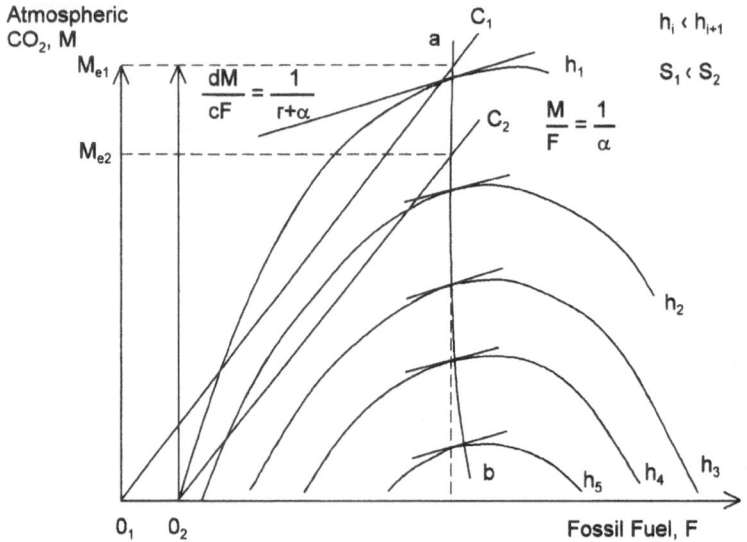

Figure 6: Case 1: A Non-fossil Substitute

The removal of CO_2 parallels this case, it changes the dynamics of CO_2 accumulation. Let S he the rate of CO_2 removal, then the state equation for the change in atmospheric CO_2 is:

$$\dot{M} = F - \alpha M - S$$

The equilibrium condition is then $F = \alpha M + S$. The change can again be represented as a shift of the vertical axis in our solution diagram to the right, a shown in Figure 6.

Case 2: Fossil Fuel Enhancement

Changes in fossil fuel productivity would result in a future production function:

f(F,M,S) = h(FS,M)

56

In equilibrium the following equation must hold along an isoquant of h:

$$-S\frac{\dfrac{\partial h(FS,M)}{\partial F}}{\dfrac{\partial h(FS,M)}{\partial M}} = S\frac{dM}{dF} = \frac{1}{r+\alpha} \tag{2}$$

This equation again can be analyzed graphically. an examination of a single isoquant in the F-M plane at two values of S, S_1 and S_2 helps explain the result. In Figure 7 the curve OC_1 represents the equation $h(FS_1,M)$. When S equals S_2, production along the curve is $h(FS_2,M)$. A higher level of technical progress compresses the isoquant toward the axis.

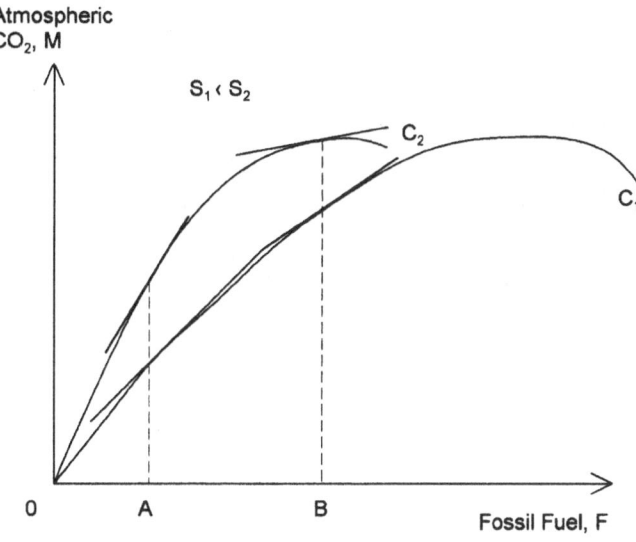

Figure 7: Case 2: Impact of Fossil Fuel Enhancement

In Figure 7 technical progress increases or decreases the slope of dM/dF along the isoquant depending on the value of F.

At A the slope increases and at B the slope decreases. Because the slopes of the isoquants may increase or decrease at any given value of F, the locus of points satisfying the rate of substitution equation (2) may move to the right or left. The change in the equilibrium level of atmospheric CO_2 is, therefore, uncertain.

An equivalent type of impact, as fossil fuel enhancement, is given by emissions scrubbing. Scrubbing CO_2 from emissions changes the dynamics of CO_2 accumulation as

$$\dot{M} = (F/S)\,\alpha M$$

The equilibrium condition is then $F = S\alpha M$. The same conclusion as in Figure 7 holds. The equivalence can easily be seen if the effects of substituting a variable F_s equal to F/S into the state equation and the production function are considered.

Case 3. Amelioration of CO_2 Impacts

Furthermore, one representation of ameliorating CO_2 impacts would change the production function as:

f(F,M,S) = h(F,M - S)

This case bears similarity to Case 1. However, in this instance the horizontal axis is raised as illustrated in Figure 8.

In Figure 8 only the isoquants for S equal to S_1 are shown, the isoquants when S equals S_2 would be moved vertically by $S_2 - S_1$ while not changing the constant CO_2 line. Consequently, the long run equilibrium level of CO_2 is raised with higher levels of technical progress.

All these cases are summarized in Table 2.

Table 2: The Effects of Finite Technical Progress		
Type of Technical Progress	Problem Modification	Sign dM/dS
1. General Technical Progress	f (F,M,S) = (F,M)	0
2. Sequestering of Carbon	No Change	0
3a. Non-fossil Substitute	f(F,M,S)h(F+S,M)	-
3b. Constant Rate of Carbon Dioxide Removal	dM/dT = F-αM-S	-
4a. Fossil Fuel Enhancement	f(F,M,S) = h(SF,M)	+-
5. Impact Amelioration	f(F,M,S)h(F,M-S)	+

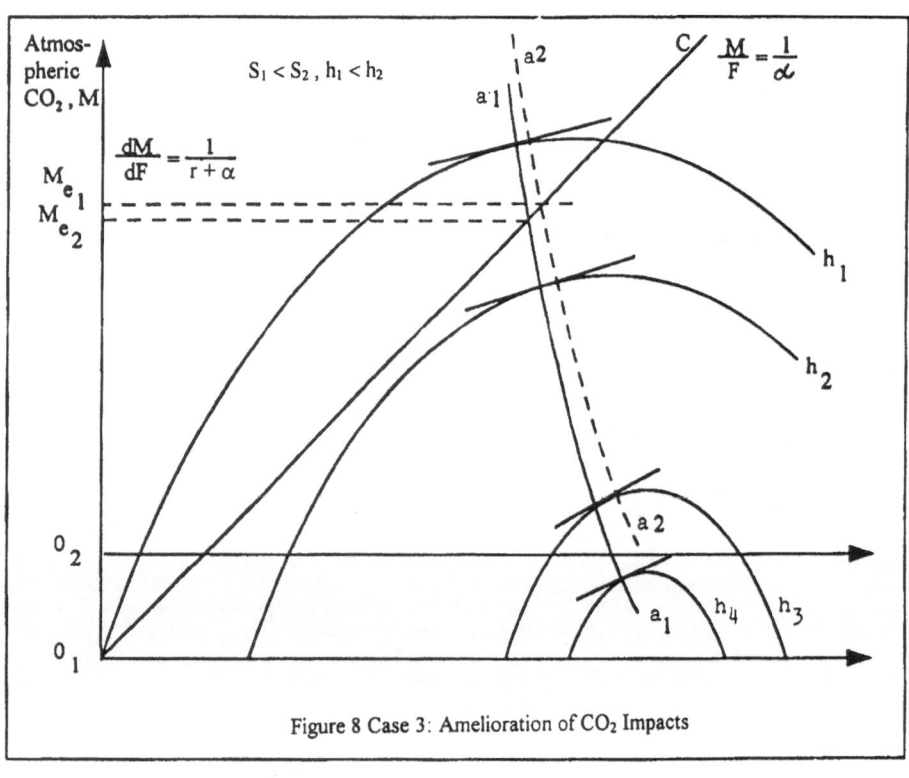

Figure 8 Case 3: Amelioration of CO_2 Impacts

60

Case 4. Exogenous Neutral Technical Progress

To examine ongoing technical progress we assume the production function to be separable
as below:

$$f(F,M,T) = h(F,M)S(t)$$

Given this production function, we restate the equilibrium condition of the price damage
equation, equation 4(25),

$$\frac{\partial q}{\partial F}\frac{dF}{dt} + \frac{\partial q}{\partial M}\frac{dM}{dt} + \frac{\partial q}{\partial S}\frac{dS}{dt} + \frac{\partial q}{\partial U'}\frac{dU'}{dt} = (r+\alpha)\, U'\frac{\partial h}{\partial F}S + U'\frac{\partial h}{\partial M}S \qquad (3)$$

In general, no equilibrium solution exists, however, when S increases exponentially with time
and the consumption elasticity of utility is constant (U takes the constant relative risk aversion
form), the problem is much simpler. dF/dt and dM/dt are zero in equilibrium. If η is
the exponential rate of increase in S, dS/dt divided by S equals γ and $\dfrac{dU'}{dt}$ divided by
U' equals $-\beta\eta$.

Then from (3) we derive the most important equation:

$$-(\partial h/\partial F)\,/\,(\partial h/\partial M) = dM/dF = \frac{1}{(r+\alpha-\eta\,(1-\beta))} \qquad (4)$$

dM/dF is defined along an isoquant of h. If the right hand side of (4) is interpreted as the
ratio of the prices of fossil fuel increases and atmospheric CO_2 decreases, the long-run costs
of CO_2 reductions are reduced by technical progress. We would expect such a cost reduction
to cause substitution away from fossil fuel use and toward reduced CO_2 levels.

The sufficiency condition on q requires that the rate of increase in utility, $\eta\,(1-\beta)$,
be less than the discount rate. This assures that the integral is finite and the maximum is well
defined. It also assures that the slope of dM/dF will be less than the slope of the line along
which αM equals F.

Using (4) we can construct Figure 9 in which the locus of solutions for (4) is drawn
for two different values of technical progress, curves a_1b_1 and a_2b_2. η_2 which defines a_2b_2
is greater than η_1 which defines a_1b_1. Figure 9 shows that the higher the rate of progress
the lower the steady state values of F and M.

We note that technical progress can be demonstrated as causing a price change which

61

induces substitution away from fossil fuel use. As we can see, higher growth moves us down the line OC onto an isoquant which has a higher level of production in the energy sector, $h(F,M)$. When the rest of the economy is growing, future output from the energy sector has a large multiplier, making higher long-run energy sector output more valuable. However, the effect on growth is also determined by the consumption elasticity of utility. The curvature of the utility function tends to level out consumption overtime. Thus it tends to reduce the weight of future consumption in a growing economy.

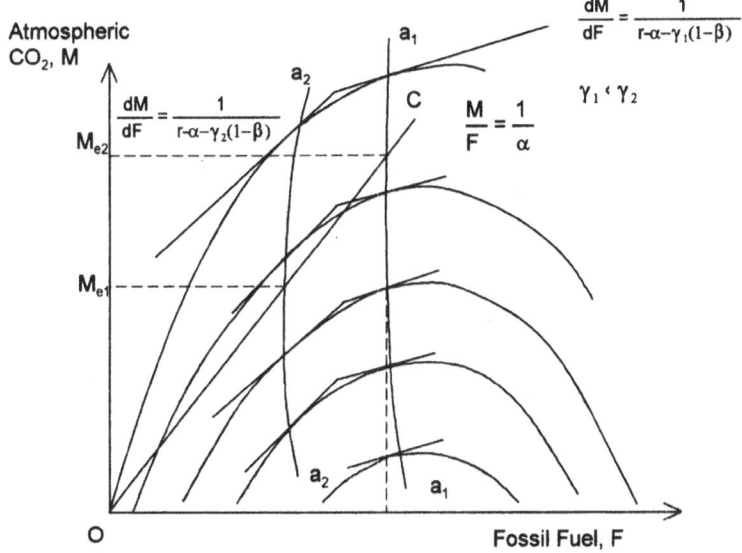

Figure 9: Exogenous, Neutral Technical Progress

2.11 EXPONENTIAL TECHNICAL PROGRESS IN A BENCHMARK MODEL

Exponential technical progress can be added to the benchmark type model developed in Section 10. In this case the production function is:

$$C = \begin{cases} F^{\epsilon} e^{\eta t}, & M \leq M_c \\ 0, & M > M_c \end{cases} \tag{1}$$

By looking at the rate of change in fossil fuel use and the initial fossil fuel use, equations 4(17) and 4(22), are restated

$$F = F(T) \, e^{A(T-t)} \tag{2}$$

$$F = A(M_c - M) + \alpha M \tag{3}$$

The variable A is redefined as

$$A = (r + \alpha - \eta(1-\gamma)) / (1 - \epsilon(1-\gamma)) \tag{4}$$

As we can see, the changes in F due to changes in the social discount rate, r, the fossil fuel productivity factor, ϵ, and the reabsorption rate, α, are as in the model without technical progress. Fossil fuel declines more slowly in this model because technical progress increases the productivity of fossil fuels. With technical progress what can we derive for the rate of growth of consumption (i.e. $(dC/dt)(1/C) = \Delta C$) ?

From the present until time T when equilibrium is reached, the rate of growth in consumption is

$$\Delta C = -(\epsilon(r + \alpha - \eta(1-\gamma)) / (1 - \eta(1-\gamma)) + \eta \tag{5}$$

The first term on the right is the rate of change in output from the energy sector, and the second term on the right is the rate of technical progress.

This expression can be simplified to

$$\Delta C = (\eta - \varepsilon (r + \alpha)) / (1 - \varepsilon (1 - \gamma)) \tag{6}$$

When the rate of growth in the economy is positive, increases in the consumption elasticity of utility, γ , tend to increase present energy use, and thereby decrease the growth rate of consumption. When the rate of growth in the economy is negative, γ has the opposite effect.

The results can be extended and partially generalized by taking other and more complicated forms of consumption equations.

In all cases model parameters affect the rate of change in energy use in the same fashion in the models with and without technical progress.

In summary we can conclude that increases in the new parameter, η , reduce both initial energy use and the rate of decline in energy use.

Equation (6) shows whether and how consumption increases or decreases over time. Technical progress tends to make consumption grow. Reabsorption, social discounting, and high fossil fuel productivity all encourage high initial energy use and make a consumption decline more likely.

2.12   INTERNATIONAL CO-OPERATION

International control agreements on CO_2 emissions would require many participants. Since temperature increases will impact countries in extremely varied ways, in some countries the impact may be beneficial. Thus international agreements on fossil fuel trade may be difficult to achieve (NAS, 1992). The basic premise is that critical CO_2 levels will be reached more quickly in a world without co-operation on CO_2 controls.

International co-operation on CO_2 emissions could be speeded up by acceleration of technology transfer between two countries or block of countries. If technology changes have a significant influence (or leverage) on critical CO_2 emissions technology transfer schemes could facilitate co-operative ventures that accelerate the innovation and diffusion of technologies for enhancing global welfare.

If CO_2 emissions and possible global warming are perceived as a threat to their survival, individual countries or a group of countries may wish to unilaterally alter the international use of fossil fuels through export and import controls such as taxes and subsidies. The general relation between global pollution and international trade has only recently received considerable attention, and the suggestion to extend the mechanism of tradeable permits to global issues, such as CO_2, has been followed up as part of the official US negotiation position on climate change (*Economist*, 1990). But the greatest part of work so far deals with local rather than transnational pollutants and is not particularly applicable to CO_2 (Gottinger, 1994).

Within the class of models previously suggested we now consider the difficulties in the control of carbon dioxide emissions.

We set out from the benchmark (step) model and compare two cases: co-operative and non-co-operative, given by subscripts 'CO' and 'NC' respectively. For example, let F_{CO} refer to world emissions in the co-operative case, F_{NC} to world emissions in the non-co-operative case. q_{CO} refers to the tax or negative adjoint variable in the co-operative case, and T_{CO} to the time at which M equals M_c in the co-operative case. In the model two world regions, 1 and 2, exist. There are n and m factories in the two world regions. Let f be the production function of a single factory. The regions use F_1 and F_2 units of energy, respectively. The factories have decreasing returns and energy is used efficiently. Factories in Region 1 use F_1/n units of energy and factories in Region 2 use F_2/m units of energy. We assume a maximum level of energy use per factory, F_m, which is greater than $\alpha M_c/(n + m)$.

Accordingly, we define the consumption in each region:

$$C_1 = \begin{cases} nf(F_1/n) , & M \le M_c \\ 0, & M > M_c \end{cases} \text{ and}$$

$$C_2 = \begin{cases} mf(F_2/m) , & M \le M_c \\ 0, & M > M_c \end{cases} \tag{1}$$

Co-operation is "complete" and the countries maximize the discounted sum of the consumption in the two regions being fully aware of future CO_2 effects.

The following setup describes the problem:

$$\underset{F_1, F_2}{Max} \quad \int_0^\infty e^{-rt}(C_1 + C_2) \, dt \tag{2}$$

$$dM/dt = F_1 + F_2 - \alpha M \tag{3}$$

Not surprisingly, the solution is similar to the pure step model. The negative of the adjoint, q_{co}, rises at an exponential rate of $(r + \alpha)$ to equilibrium at $f'(\alpha M_c/(m+n))$. The marginal product of energy in every factory equals the negative of the adjoint variable, f' equals q_{co}. The inverse of this function, say $\phi(q)$, determines F_{1co} and F_{2co}:

$$F_{1co}(t) = n\phi(q_{co}(t)) , F_{2co} = m\phi(q_{co}(t)) \tag{4}$$

T_{co} is defined by

$$\int_0^{T_{co}} e^{\alpha(T_{co}-t)} (n+m) \, \phi\{f'[\frac{\alpha M_c}{m+n}] \, e^{(r+\alpha)(T_{co}-t)} \, dt - M_o e^{-\alpha T_{co}} = M_c \tag{5}$$

The non-co-operative case differs considerably. Region 1 recognizes the future CO_2 impacts and control emissions, but Region 2 is unaware of or ignores the impacts until M equals M_c. Region 2 initially burns fuel at a level mF_m such that $f'(F_m)$ equals 0.

For simplicity we assume that when M equals M_c co-operation begins and that F_{1NC}/n equals F_{1NC}/m and $F_{1NC} + F_{2NC}$ equals αM_c. Region 1 solves the following problem:

$$\underset{F_1}{Max} \int_O^\infty e^{-rt} C_1 dt \tag{6}$$

$$\frac{dM}{dt} = \begin{cases} F_1 + mF_m - \alpha M, & M \le M_c \\ O, & M > M_c \end{cases} \tag{7}$$

q_n rises exponentially at the rate $(r + \alpha)$ until equilibrium at $f'(\alpha M_c/(m + n))$. In Region 1 only, the marginal products of factories are held at q_n and emissions of these factories are equal to $\phi(q)$. F_{1NC} and F_{2NC} for $M < M_c$ are:

$$\begin{aligned} F_{1NC} &= n\phi(q_n(t)) \\ F_{2NC} &= mF_m \end{aligned} \tag{8}$$

T_{NC} is then given by

$$\int_O^{T_{NC}} e^{-\alpha(T_{NC}-t)} \{ n\phi [f'(\frac{\alpha M_c}{m+n}) e^{-(r+\alpha)(T_{NC}-t)}] + F_m \} dt - M_O e^{-\alpha T_{NC}} = M_c \tag{9}$$

Thus we could state the theorem.

Theorem

A critical CO_2 emissions level is always reached in less time in the non-co-operative case than in the co-operative case. T_{NC} is less than T_{CO}.

In order to prove this result, we restate (5) and (9) respectively, as:

$$\int_O^{T_{CO}} e^{-\alpha t} \{ (n+m) \phi [f'[\frac{\alpha M_c}{m+n}] e^{-(r+\alpha)t}] - \alpha M_O \} dt + M_O = M_c \tag{10}$$

and

$$\int_O^{T_{NC}} e^{-\alpha t} \{ n\phi [f^1 [\frac{\alpha M_c}{m+n}] e^{-(r+\alpha)t}] + mF_m - \alpha M_O \} dt + M_O = M_c \tag{11}$$

Essentially the integration is now backwards in time. Because F_m is greater than $m\phi[f'(\alpha M_c/(m+n)) \exp(-(r+\alpha)t)]$ and the other terms of the integrands are equal, at each instant the integrand in (11) is greater than the integrand in (10). Since the integrand is always larger in (11), the integral must equal $M_c - M_O$ in a shorter time.

We have shown previously that $q_{NC}(T_{NC})$ equals $q_{co}(T_{CO})$ equals $f'(\alpha M_c/(m + n))$.

If we single out the specific Region 1 and let T_{NC} be less than T_{CO}:

$$q_{NC}(t) = q_{NC}(T_{NC}) e^{-(r+\alpha)(T_{NC}-t)} > q_{CO}(T_{CO}) e^{-(r+\alpha)(T_{CO}-t)} = q_{CO}(t) \tag{12}$$

$$n\phi(q_{NC}(t)) = F_{1NC} < F_{1CO} = n\phi(q_{CO}(t)) \tag{13}$$

The emissions of the concerned region are lower in the non-co-operative case than they are in the co-operative case.

T_{NC} less than T_{CO} appears to imply that world fossil fuel use in the non-co-operative case is always higher than in the co-operative case; however, this is not necessarily true. There could be special situations identified by specific parametric configurations of the production function and other parameters, in which world emissions in the non-co-operative case are actually lower than in the co-operative case for a short initial period.

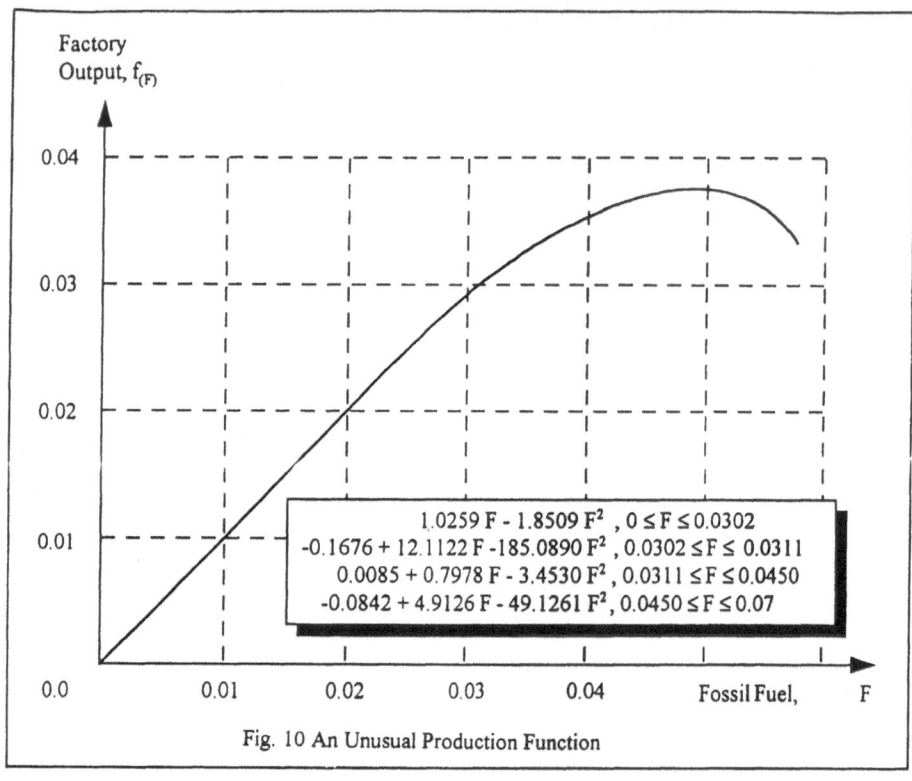

Fig. 10 An Unusual Production Function

Figure 10 illustrates a production function which can cause such unusual behaviour.

This production function is a combination of the quadratic curves listed in the figure. The first derivative is continuous and the second derivative is always negative. Other parameters in our example are M_0 equals 1.0, M_c equals 1.4, α equals 0.01, r equals 0.08, and n and m equal 1. Then F_{1NC} (T_{NC}) and F_{1CO} (T_{CO}) equal $\alpha M_c/(m+n)$ which equals 0.007. q_{NC} (T_{NC}) and q_{CO} (T_{CO}) equal f' ($\alpha M_c/(m+n)$) which equals 1. From (12) $q_{NC}^{(t)}$ equals exp ($-(r+\alpha)(T_{NC}-t)$) and $q_{CO}^{(t)}$ equals exp ($-(r+\alpha)(T_{CO}-t)$).

The marginal production function f'; is a series of linear equations. Therefore, the inverse $\varnothing$ (q(t)) is also a series of linear equations. These curves are shown in Figure 11.

Substituting (q(t)) into (11) and (12) T_{NC} equals 6 and T_{CO} equals 8. The $q_{NC}(0)$ equals

70

0.58275 and $q_{co}(0)$ equals 0.48675. From figure 11, ϕ ($q_{NC}(0)$) equals 0.03114 and ϕ ($q_{co}(0)$) equals 0.04504. The total emissions at time 0 in the base case for these parameters are 2 ϕ ($q_{co}(0)$) equals 0.09008. The total emissions in the non-cooperative case equal ϕ ($q_{NC}^{(0)}$) or 0.03114 plus ϕ (0) or 0.05000 or in total 0.08114. Emissions in the cooperative case are about ten percent higher than in the non-cooperative case.

Examining figure 11 the reason for this emission pattern is evident. The graph of the inverse function, ϕ (q), is very steep for values of q just less than 0.6, and the price elasticity of demand for fossil fuels is very high. Because $q_{NC}^{(0)}$ and $q_{co}^{(0)}$ are in this region, the relatively small increase in q results in a very large drop in fossil fuel use.

In summarizing the results, if co-operation is not feasible, regions concerned about carbon dioxide will lower emissions to compensate for regions not concerned about carbon dioxide, however, the maximum level of carbon dioxide concentration will be reached later with co-operation than without co-operation.

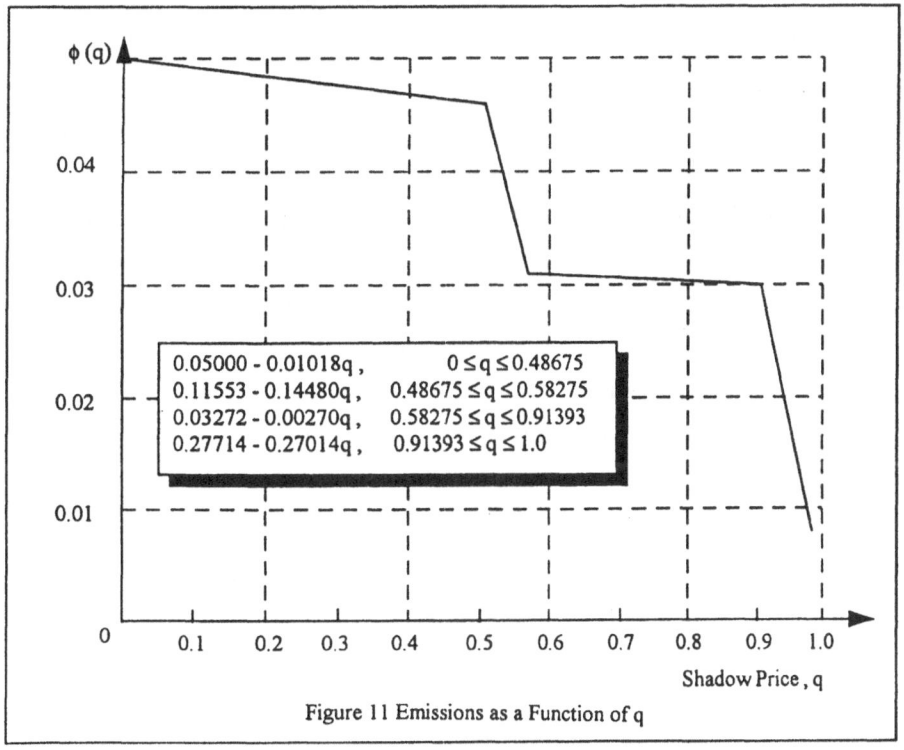

Figure 11 Emissions as a Function of q

2.13 CONCLUSIONS AND PERSPECTIVES

The choice of simple models proposed so far have been examined under three major issues: technical progress, international co-operation and structural uncertainty. We show that for this class of models depending on the assumptions regarding technical progress, the optimal steady state CO_2 concentration may rise or fall with increases in the steady state level of progress. More specifically, an improved substitute for fossil fuels always reduces the long-run level of CO_2; while an improvement in fossil fuel productivity may increase or decrease the level of CO_2.

Also we show that the solutions of a model with neutral, constant and ongoing technical progress and the general (static) model are very similar. In the model with technical progress, higher levels of technical progress lead to lower long-run optimal levels of CO_2.

By examining two cases of international co-operation in controlling CO_2 accumulation, we first considered a reference case of complete co-operation between two regions in maximizing consumption with full awareness of the CO_2 problem. This case is compared with a situation in which no co-operation takes place until a critical CO_2 level is reached. A most interesting finding is that in the non-co-operative situation the critical level is reached sooner, even though one region (concerned about CO_2) always emits less carbon than in the co-operative case. A non-intuitive situation can occur in which initial emissions of the two regions are lower in the non-co-operative case than in the co-operative case.

Some final comments should be devoted to the general philosophy of modelling complex, highly interactive energy-environmental situations. Our task has been ambitious, namely to exhibit single type models of optimal control that are able to identify major structural parameters of the CO_2 problem as seen from an economist's perspective.

In models which optimize over an infinite horizon, future effects may change current policies, and feedback effects are always of importance. However, in these same models feedback effects may make solution much more difficult. In this class of optimizing models inclusion of feedback effects usually lowers the optimal initial use of fossil fuels. The long-run changes in fossil fuel use due to feedback effects are more uncertain and dependent on the model.

Given the uncertainty in the severity and timing of feedback effects, the sensitivity of individual models to variations in feedback effects is of much interest. In this class of models we find that current optimal fossil fuel use is significantly affected by different critical levels of CO_2.

Including optimization in models expands their applicability but may cause analytic problems and controversy. As with many social policy problems, involving welfare judgements, an acceptable objective function for CO_2 control problems is difficult to define.

Any definition will seem both inadequate and overly precise and certainly will be controversial. This may be the reason we will find very few optimizing models in this area. However, such a model will not hide useful results otherwise obtained and by a carefully done sensitivity analysis may add many new insights not otherwise obtained. I concur with R. E. Lucas (1987, chap. 2) that "useful policy discussions are ultimately based on models".

Optimization raises several new issues in these models. For example, pollution impoverishes but technical progress enriches the future.

The models show that curvature of the utility function, determined by the consumption elasticity of utility in our models, tends to smooth or even out wealth over time. Without an objective function being stated, the importance of this redistribution effect in determining fossil fuel use policy can not be examined.

APPENDIX A

Existence and Stability of Equilibrium

In this appendix we show that the equilibrium solution of Section 4 holds providing in detail the assumptions which assure our solution holds.

Several assumptions assure that a solution to 4(25') exists when F equals αM. As with the sufficiency conditions, the general existence proof is only applicable if the production function has continuous first derivatives. Continuous first derivatives assure that the left-hand side (lhs) of 4(25') is continuous except in any region where $\partial f/\partial M$ equals zero. Experience has shown that CO_2 accumulation has caused little damage to date and that energy is valuable in production; therefore, we assume that at zero energy use and zero atmospheric CO_2, the marginal product of energy use is greater in magnitude than the marginal harm due to an increase in CO_2. It follows that the lhs of 4(25') is very large when F and αM equal zero.

Existence of the equilibrium is then assured if at some level of atmospheric CO_2 concentration the marginal product of energy use is less in magnitude than the marginal harm due to an increase in atmospheric CO_2 i.e. the lhs of 4(25') is less than one. This can be assured by two reasonable assumptions:

$$\frac{\partial f(0,M)}{\partial F} \leq \frac{\partial f(0,0)}{\partial F} \tag{A1}$$

$$\frac{\partial f(\alpha M,M)}{\partial M} > \frac{\partial f(0,0)}{\partial F} \tag{A2}$$

as M goes to infinity.

The first assumption is that the marginal product of fossil fuels is greatest at zero fossil fuel use and atmospheric CO_2 accumulation.

The second condition states that, as atmospheric CO_2 increases, at some point the marginal harm is greater than the maximum marginal product of fossil fuels. More stringent assumptions, which assure that equations (A1) and (A2) are satisfied, are that the second

74

partial derivative of f with respect to F and M are negative and that the marginal harm due to CO_2 goes to infinity.

By taking the total differential of 4(25') we can show that, if the second partial derivative of f with respect to F and M are negative, we are also assured that curve ab in Figure 1 slopes downward as shown.

Although we have discussed the conditions which show that the equilibrium is optimal and exists, we still have to show that an optimal path to the equilibrium exists. If the equilibrium is stable or, in other words, a saddle point, a unique shadow price exists for each level of CO_2 concentration which if chosen initially will cause convergence to the equilibrium along the optimal path. If the equilibrium is unstable, the equilibrium is maintained when CO_2 is initially at the equilibrium level; but no optimal path to the equilibrium exists if CO_2 starts at any other level.

Our stability analysis follows Kamien and Schwartz (1981, p 160). The proof of stability depends on the particular structure of our simple model. When F is greater than zero, 4(15) is an equation of the form $G(F, M, q) = 0$. We earlier assumed that we could express F as a function of M and q. By the implicit function theorem, this function for G exists if the partial derivative of G with respect to F does not equal zero for any F, M, and q that satisfy 4(15). The partial derivative of f with respect to F and the derivative of U with respect to C have already been assumed to be non-zero for values of F in the range of interest, therefore, the function for F in terms of q and M exists.

The next step in the stability analysis is to linearize equations 4(24) and 4(25) about the equilibrium. If the equilibrium values of M and q are designated $\overline{M}$ and $\overline{q}$, then 4(24) and 4(25) can be approximated.

$$-\left[\alpha + \frac{U_{fm}}{U_{ff}}\right](M-\overline{M}) + \frac{1}{U_{ff}}(q-\overline{q}) = 0 \tag{A3}$$

$$-\left[\frac{U_{fm}^2 - U_{ff}U_{mm}}{U_{ff}}\right](M-\overline{M}) + \left[r + \left(\alpha + \frac{U_{fm}}{U_{ff}}\right)\right](q-\overline{q}) = 0 \tag{A4}$$

A linear system, as in (A3) and (A4) has a characteristic equation. If the roots of the equation are real and of opposite signs, the equilibrium is a saddle point and the solution is stable. We state the derived stability condition without further discussion of the characteristic equation. For further details we refer to the discussion of stability analysis in Kamien and

Schwartz (1981). If the following inequality holds, the equilibrium is stable:

$$U_{mm} + (r\alpha + \alpha^2)U_{ff} + (r + 2\alpha)U_{fm} < 0$$

In our discussion of sufficiency we assumed that U_{ff} and U_{mm} were both negative, we again make these assumptions. The final assumption is that U_{fm} is less than or equal to zero and consequently that the equilibrium is stable. This assumption can only be true if the second partial derivative of production with respect to fossil fuel use and atmospheric CO_2 is less than or equal to zero, see equation 4(20). This assumption makes sure that the marginal product of fossil fuel use does not increase with increases in atmospheric CO_2. Note, the relation in (A3) may hold even if U_{fm} is greater than zero.

In our analysis of the equilibrium we have found two key conditions. First, for the equilibrium itself to be optimal, the second partial derivatives of production with respect to both fossil fuel use and atmospheric carbon dioxide must be negative, and the condition in 4(23) must hold. Second, for a simple and general proof that an optimal path to the equilibrium exists, U_{fm} in 4(20) must be negative. This implies that the second partial derivative of production with respect to fossil fuel use and atmospheric carbon dioxide must be negative. We must emphasize that these conditions are not necessary, they only allow easy proofs of sufficiency and stability. If these conditions hold and M is initially lower than the equilibrium, CO_2 increases, the shadow price increases, and fossil fuel use decreases with time until equilibrium.

APPENDIX B

Proposition

If f in Section 5(1) has no maximum or the maximum occurs at F greater than αM_c and M(0) is less than M_c there is a unique finite time T at which M equals M_c.

Proof. If it exists, the time at which M equals M_c, T, is defined by:

$$M_c = \int_0^T e^{-\alpha(T-t)} g(q(t)) dt + e^{-\alpha T} M_o \tag{B1}$$

When M equals M_c or t is greater than or equal to T, q equals $U'f'(\alpha M_c)$.

From 5(3), when M is less than M_c or t is less than T, q equals $q_o \exp(r + \alpha)t$ where q_o is the value of q at t equals 0. Continuity of q requires that, for an optimal path:

$$U'f'(\alpha M_c) = q_o e^{(r+\alpha)T} \tag{B2}$$

If a q_o and T satisfying (B1) and (B2) can be found the paths of q, F and M are all determined.

(B2) can be solved for q_o in terms of T. (B1) can then be restated as an equation in which T is the single unknown variable. After rearranging terms:

$$\int_0^T e^{-\alpha T} (g(U'f'(\alpha M_c) e^{-(r+\alpha)(T-t)}) - \alpha M_c) dt - e^{-\alpha T}(M_c - M_o) = 0 \tag{B3}$$

At T equals zero the integral in (B3) is zero and at T equals infinity the integral is infinity. The function is continuous, therefore, a solution exists and is finite. The derivative of the left-hand side with respect to T is positive, therefore the solution is unique.   q.e.d.

REFERENCES

Chiarella, C., "Optimal Depletion of a Nonrenewable Resource when Technological Progress is Endogenous", in Kemp, M. C. and Long, N. V. (eds), *Exhaustible Resources, Optimality and Trade*, North Holland, Amsterdam, 1980, pp. 81-93.

Conrad, J. M., and Clark C. W., *Natural Resource Economics*, Cambridge University Press: Cambridge, 1987.

Deshmukh, S. D. and Pliska, S. R., "Optimal Consumption of Non-renewable Resources with Stochastic Discoveries and a Random Environment", *Review of Economic Studies*, 50, July 1983, pp. 543-554.

Deshmukh, S. D. and Pliska, S. R., "Optimal Consumption and Exploration of Non-renewable Resources with Uncertainty", *Econometrica* 48(1), 1980, pp. 177-200.

The Economist, "Greenhouse Economics", "Trading Places", July 7, 1990, 19-22, 46-47.

The Energy Journal, *Special Issue on Global Warming*, 12(1), 1991, pp. 1-196.

Edmonds, J. and Reilly, J., "Global Energy and CO_2 to the Year 2050", *Energy Journal* 4(3), 1983, pp. 21-47.

Fisher, A. C., *Resource and Environmental Economics*, Cambridge University Press: Cambridge, 1981.

Gottinger, H. W., "Some Policy Issues of Greenhouse Gas Economics", Nordic Journal of Environmental Economics 5, 1994, pp. 21-31.

Halkin, H., "Necessary Conditions for Optimal Control Problems with Infinite Horizons". *Econometrica*, 42, 1974, 267-272.

Kamien, M. I. and Schwarz, N. L., *Dynamic Optimization*, North Holland: Amsterdam, New York, 1981, second edition, 1991.

Kemp, M., "How to Eat a Cake of Unknown Size" in Kemp, M. C. (ed.), *Three Topics in the Theory of International Trade*, North Holland: Amsterdam, 1976.

Krelle, W., Wirtschaftswachstum bei Erhaltung der Umweltqualitat (Economic Growth by Maintaining Environmental Quality), in R. Henn (ed.), *Technologie, Wachstum und Beschaftigung*, Springer: Heidelberg, 1987, pp. 757-778.

Lucas, R. E., *Models of Business Cycles*, Yrjo Johnsson Lectures, Basil Blackwell: Oxford, 1987.

Manne, A. S., "ETA-MACRO: A Model of Energy-Economy Interactions" in Charles J Hitch (ed.), *Modelling Energy-Economy Interactions*, Resources for the Future,

Washington   D.C., 1977.

Manne, A. S. and Richels, R. G., "The Costs of Reducing U.S. CO_2 Emissions", *Energy Journal*, 11(4), 1990, 69-78, Further Sensitivity Analysis.

Manne, A. S., and Richels, R. G., Buying Greenhouse Insurance, MIT Press: Cambridge, Mass. 1992a.

National Academy of Sciences (NAS), Policy Implications of Greenhouse Warming, Committee on Science, Engineering, and Public Policy, National Academy Press: Washington D.C., 1992.

National Research Council (NRC), Carbon Dioxide Assessment Committee. Nierenberg, W. A. (ed.) *Changing Climate*, National Academy Press: Washington D. C., 1983.

Nordhaus, W. D., "Thinking about Carbon Dioxide: Theoretical and Empirical Aspects of Optimal Control Strategies". Cowles Foundation Discussion Paper No 565, Yale University, New Haven, Conn. Oct 1980.

Nordhaus, W. D., "Rolling the 'Dice': An Optimal Transition Path for Controlling Greenhouse Gases", *Resource and Energy Economics* 15, 1993, 27-50.

Nordhaus, W. D., "*Managing the Global Commons, The Economics of Climate Change*, MIT Press, Cambridge, Mass., 1994.

Okken, P. A., Swart, R. J. and Swerver, S. (eds), *Climate and Energy*, Kluwer Academic Publishers: Dordrecht, 1989.

Peck, S. C., and T. J. Teisberg, "CETA: A Model for Carbon Emissions Trajectory Assessment", *Energy Journal* 13 (1), 1991, 55-77.

Perry, A. M. et al, "Energy Supply and Demand Implications of CO_2", *Energy Journal*, 13(12), 1982, pp. 991-1004.

Rose, D. J., Miller, M. M. and Agnew, C., *Global Energy Futures and CO_2 Induced Climate Change*. Report MITEL 83-015, MIT Energy Laboratory, Cambridge, Massachussetts, Nov 1983.

Rosenberg, N. J., "A Primer on Climatic Change: Mechanisms, Trends and Projections". Renewable Resources Division, Resources for the Future, Discussion Paper RR 86-04, August 1986.

Sato, R., *Theory of Technical Change and Economic Invariance*, Ch. 4., Academic Press: New York, 1981.

Yohe, C. W., "The Effects of Changes in Expected Near-term Fossil Fuel Prices on Long-term Energy and Carbon Dioxide Projections", *Resources and Energy*, 6, 1984, pp. 1-20.

Footnotes

[1]   In finite horizon problems with a structure similar to (1) - (9) an additional necessary condition determines the value of the adjoint variable, ρ or ϕ, at the terminal time. Halkin (1974) showed that the simple condition on the adjoint variable in the finite horizon case does not extend to the infinite horizon case. However, there exist conditions on the shape of the function to be maximized and the state equation which in combination with (14), (15) and (16) are sufficient to determine an optimum.

CHAPTER 3

UNCERTAINTY IN ECONOMIC MODELS OF OPTIMAL ENERGY USE

3.1   INTRODUCTION

In the economics of non-renewable resources the problem of eating the cake when its size is unknown constitutes an essential feature of uncertainty. Such a model was proposed by Gilbert (1979) and he obtained results that were both illuminating and suggestive. The problem of consuming an exhaustible resource when there is ignorance about its size is analytically equivalent to the more recent problem of carbon dioxide (CO_2), that is generating fossil fuel emissions when the critical CO_2 budget (where irreversibly the damage becomes infinite) is uncertain. It is also interesting to note that uncertainty reducing activities such as exploration in the former problem correspond to the use of research in the latter.

We will convert Gilbert's model to the new problem context and show, in a social planning framework, that the treatment of uncertainty has significant impacts on policy choices regarding global CO_2 emission limits.

We will support such conclusions by a numerical treatment of an optimal control model on fossil fuel use where uncertainty is a major subject of sensitivity analysis.

Existing energy-economy-environmental (EEE) models suffer from poor data, indeterminate structure, and a frequent lack of attention to the consequences of uncertainty. The factors linking energy activities to their environmental effects are known only imprecisely. EEE models rely on behavioral assumptions that are widely questioned, and on parameters that can vary substantially from one model or data source to the next.

Most EEE models, including those well-established and highly used, conceal this uncertainty behind a blanket of output detail: a profusion of fuel prices and quantities, sectoral disaggregation, regional detail, growth rates and target figures, which often steer the analysis toward a desired conclusion. Unfortunately, this complexity rarely contributes to a resolution of uncertainty, and may serve only to increase the error and expense.

Concerning models of possible greenhouse effects and CO_2 emissions, uncertainty analysis assumes many facets. It involves:

(i)   changes in climate to be expected

(ii)   impact of climate change

(iii)   costs of adapting to climate change

We distinguish between uncertainty about occurrences of events and impacts. Policy uncertainty is also of great concern (Ingham and Ulph, 1991; Henrion and Morgan, 1990). Shall we act now and learn later, or learn now and act later? For the CO_2 problem some argue that it is premature to think about doing other than intensive research, others claim that the risks of waiting are simply too great. What is the value of reducing scientific uncertainty? Scenario analysis only provides an indirect treatment of uncertainty, all uncertainties are resolved prior to decision-making.

As part of our optimal control model in Chapter 2, we consider structural uncertainty as affecting the model's parameters. Structural uncertainty is an inherent property of complex EEE models. It relates to the stochastic nature of optimal control models (Holly and Hallet, 1989), adapted to our specific needs.

In such models the treatment of uncertainty of critical parameters by the use of expected values is common, but we show that anything less than a full treatment of uncertainty can lead to a biased calculation of the optimal present fossil fuel use.

This paper shows that the results from studies of the optimal use of a resource which is in limited supply can be applied to the CO_2 problem.

Then, using numerical examples, we show that an inappropriate treatment of uncertainty can lead to significantly higher than optimal estimates of the desirable level of fossil fuel use.

We note that the determinations of optimal fossil fuel use are similar when there is an uncertain, critical atmospheric CO_2 level and when the fossil fuel resource is uncertain. In fact, when there is zero absorption of atmospheric CO_2, the problems are identical. When absorption is greater than zero, the results of research on uncertain, limited, natural resources are often applicable to the CO_2 problem with only minor changes in definitions.

By making use of the results of Gilbert (1979) the theorems used here are so similar to those of Gilbert that the proofs will not be presented.

We limit the treatment of uncertainty to this case, other forms of uncertainty linked to value of information, irreversibility, learning and optimal stopping notwithstanding. (Chichilnisky and Heal, 1993; Kolstad, 1992: Peck and Teisberg, 1993; Chapters 9 and 10.

3.2 STRUCTURAL UNCERTAINTY IN A SIMPLE AGGREGATE EEE MODEL

To demonstrate how uncertainty naturally enters our simple aggregate step model in Chapter 2, we assume that the present level of CO_2 is $M_o = M(0)$ and that n distinct (possible) carbon dioxide levels, M_i exist. The prior probability that the critical CO_2 level M_c equals M_i is π_i. J_i is the maximum expected value of future fossil fuel consumption when the current level of carbon dioxide in the atmosphere is M_i. π_{ij} is the updated probability that M_c equals M_j given that M_i has been reached and is not the critical level. E denotes an expectation, r is the discount rate. Furthermore, U (C) is a utility stream of consumption, production good C depends on the use of fossil fuels, $C = f(F)$, but production drops to zero if a critical CO_2 concentration is exceeded. J_i is defined by

$$J_i = Max \ E \left[\int_0^\infty e^{-rt} U(C) dt \right] \tag{1}$$

such that

$$C = \begin{cases} f(F), M \le M_c \\ 0, M > M_c \end{cases} \tag{2}$$

$$M(0) = M_i \tag{3}$$

$$\pi_{ij} = P(M_j = M_c \ given \ that \ M_i \ne M_c)$$
$$= \begin{cases} \pi_k / \sum_{j=i+1}^{n} \pi_j, k = i+1, ..., \\ 0, otherwise \end{cases} \tag{4}$$

Let T_i be the time to move between CO_2 levels M_i and M_{i+1}.

Define the emission rate

$$\int_0^{T_i} e^{-\alpha(T_i-t)}F(t)dt = M_{i+1} - e^{-\alpha T_i}M_i \qquad (5)$$

with α being the CO_2 reabsorption coefficient, i.e. the fraction of CO_2 reabsorbed by the earth through various mechanisms.

Then J(M_i, M_{i+1}, T_i), the value collected moving from a CO_2 concentration of M_i to M_{i+1} in time T_i is defined as

$$J(M_i, M_{i+1}, T_i) = \underset{F}{Max} \int_0^{T_i} e^{-rt}U(C)dt \qquad (6)$$

such that

$$C = f(F) \qquad (7)$$

$$\frac{dM}{dt} = F - \alpha M \qquad (8)$$

that is, CO_2 leaves the atmosphere naturally at a rate proportional to the CO_2 concentration. Furthermore,

$$M(0) = M_i \ and \ M(T_i) = M_{i+1} \qquad (9)$$

Gilbert's analysis (1979) is pertinent to our model. In order to facilitate optimization of J_i we could establish the algorithm

84

$$J_i = \underset{T_i}{Max} \left\{ J(M_i, M_{i+1}, T_i) + e^{-rT} \left[p_{i,i+1} \frac{U(C(\alpha M_{i+1}))}{r} + (1 - p_{i,i+1}) J_{i+1} \right] \right\} \quad (10)$$

The solution algorithm (10) requires maximization over T_i of three terms. The first is the value gained while raising the CO_2 level from M_i to M_{i+1}. The second is the discounted value of consuming fuel at a rate that maintains the CO_2 level at M_{i+1} weighted by the probability that M_{i+1} is the critical CO_2 level. The third is the discounted, expected value of raising the CO_2 concentration above M_{i+1} weighted by the probability that M_{i+1} is not the critical level.

A further result is that a certainty equivalent critical level of carbon dioxide, M_{ce}, exists. When used in calculations of optimal emissions, M_{ce} is a certainty equivalent level of CO_2 which produces the same initial emissions as the algorithm in (10).

But within this framework of treating structural uncertainty we can show that M_{ce} is less than $E(M_c)$. It follows that the calculated optimal current fossil fuel use, when uncertainties are fully treated, (FTu) in terms of certainty equivalence formulation, is lower than the optimal current fossil fuel use, when expected values are treated as certain values (EC).

Looking at particular examples, we compare two different treatments of uncertainty. In the 'expected case', the expected value of uncertain parameters is used in all calculations, and other aspects of uncertainty are ignored. In the 'base case', uncertainty is fully treated, that is, the certainty equivalence formulation is applied. If the step model of CO_2 impacts is used, Chapter 2, then changes in present fossil fuel use under the two treatments of uncertainty and a wide variety of parameter settings can be calculated. In the expected case suppose M_c equals M_1 with probability π and M_2 with probability $(1 - \pi)$. If the critical CO_2 level is assumed equal to the expected critical level

$$M_c = \pi M_1 + (1 - \pi) M_2$$

The problem is now 'certain', the present emissions level can be found, as in the step model. In the second approach to uncertainty, we make use of the probability estimates and the algorithm developed in (10). In this case, the atmospheric CO_2 certainly can increase from

M_0 to M_1. $J(M_0, M_1, T)$, the value gained in the transition is certain, and calculated according to (10).

3.3 EXAMPLE IN A SPECIFIED MODEL

We now show on the basis of the previous results that the way in which uncertainty is treated in such models can significantly alter the calculated response to CO_2. For reasons of simplicity, let us first assume that no (observable) CO_2 impacts occur and no information is gathered about CO_2 impacts until the CO_2 level reaches a threshold M_a. At that time all uncertainty regarding the future impacts of CO_2 is resolved.

This simplifies the previous model because all learning occurs at a single CO_2 level rather than at several levels.

In this form, the problem can be stated as

$$J_O = \frac{Max}{T}\{J(M_O, M_a, T) + e^{-rT} E(J_a)\} \tag{1}$$

T is the time at which the CO_2 level equals M_a. $E(J_a)$ is the expected utility gain after the CO_2 level reaches M_a and uncertainty is resolved. J_a does not depend on T. Utility is

specified by $U(C) = \begin{cases} \dfrac{C^{1-\gamma}}{1-\gamma}, \gamma \neq 1 \\ \ln C, \gamma = 1 \end{cases}$, up to the critical level γ being the consumption

elasticity of utility, production is specified as $f(F) = \lambda F^{\delta}$, with λ a scaling factor and δ the fossil fuel productivity factor.

From the standard necessary conditions, Chapter 2, we derive

$$F(t) = F_O e^{-At} \text{ for } t \leq T, \text{ where } A = (r + \alpha)/(1 - \delta(1 - \gamma)) \tag{2}$$

Using 2(5) and (2) we determine F_O as a function of T:

$$F_O=(M_a e^{\alpha T}-M_O)(A-\alpha)/(1-e^{-(A-\alpha)T}) \tag{3}$$

The utility gained while increasing the CO_2 level from M_O to M_a, $J(M_O, M_a, T)$ is certain and a function of T:

$$J(M_O,M_a,T)=\int_O^T e^{-rt}u(C)dt=[F_O^{\delta(1-\gamma)}/(1-\gamma)(A-\alpha)](1-e^{-(A-\alpha)T}) \tag{4}$$

The solution algorithm from equation (1) then takes the form

$$J_O=\frac{Max}{T}\{[F_O^{\delta(1-\gamma)}/(1-\gamma)(A-\alpha)](1-e^{-(A-\alpha)T})+e^{-rT}E(J_a)\} \tag{5}$$

To determine the maximum, the derivative with respect to T is set to zero, recalling that F_O is a function of T:

$$dJ_O/dT = 0 \tag{6}$$

Thus, for (5)

$$F_O^{\delta(1-\gamma)}\{[\delta/(A-\alpha)F_O](1-e^{-(A-\alpha)T})(dF_O/dT)+$$
$$[1/(1-\gamma)]e^{-(A-\alpha)T}\}-re^{-rT}E(J_a)=0 \tag{7}$$

If, from (3), the derivative of F_O with respect to T is substituted into (7), we find

$$F_O^{\delta(1-\gamma)}\{[\delta\alpha M_d/F_O]e^{\alpha T}+[1-\delta(1-\gamma)/(1-\gamma)]e^{-(A-\alpha)T}\}-re^{-rT}E(J_a)=0 \tag{8}$$

Since we can define

$$F_a=F_O e^{-AT} \tag{9}$$

we can simplify (8), as

$$[F_a^{\delta(1-\gamma)}/r][\delta\alpha M_d/F_a)+(1-\delta(1-\gamma)/(1-\gamma))]=E(J_a) \tag{10}$$

(10) can be solved by numerical methods for any value of $E(J_a)$. It gives rise to a correct, appropriate treatment of uncertainty. Thus, T and the optimal present emissions can be found.

3.4   IMPACT OF UNCERTAINTY TREATMENT AND PARAMETER CHANGES

In comparing two different treatments of uncertainty, we compare the "expected value" case from the comprehensive treatment of uncertainty, as shown by the formulae above culminating in equation 3(10).

To show even more specific results we use our general benchmark step model and calculate changes in present fossil fuel use under the two treatments of uncertainty for a variety of parameter settings.

Let us start with the familiar framework: M_c equals M_1 with probability π and M_2 with probability $(1 - \pi)$. Let the critical CO_2 level be assumed equal to the expected critical level:

$$M_{ca} = \pi M_1 + (1 - \pi) M_2 \qquad (1)$$

where M_{ca} is the assumed certain critical level. The problem can now be treated like a deterministic problem and the results obtained in Chapter 2 directly apply.

In the second approach to uncertainty, we make use of the probability estimates and the algorithm developed in the previous section.

In this case, the CO_2 level certainly can increase from M_0 to M_1. $J(M_0, M_1, T)$, the value gained in this transition, is certain and is given by 3(4). As noted above, $E(J_a)$ represents the expected utility gathered after M_1 is reached. $E(J_a)$ is the weighted sum of the utility gained by staying at M_1 and the utility gained by going to M_2 and staying at that level of CO_2.

Let $J_s(M)$ be the utility gathered when the CO_2 level is maintained at a constant M from the present into infinity.

The following equation defines $J_s(M)$ for a general production function and for the production function F^δ, in particular:

$$J_s(M) = U(f(\alpha M, M))/r = (\alpha M)^{\delta(1-\gamma)}/r(1-\gamma) \tag{2}$$

If M_1 is the maximum level of CO_2 concentration, the utility is specified by (2) with M equals M_1. If M_2 is the maximum level of CO_2 concentration, $J(M_1, M_2, T^*)$ defined by 3(4), is collected going from M_1 to M_2, and $J_s(M_2)$ is collected after M_2 is reached. T^* is the optimal time to go from M_1 to M_2. With M_0 equal to M_1 and M_c equal to M_2, we can determine T^*, by a simple observation[1]

$$[\alpha M_2/(A-\alpha)M_1][e^{(A-\alpha)T_*} - 1] + e^{-\alpha T_*} - 1 = 0 \tag{3}$$

After T^* being defined by (3), $E(J_a)$ can now be calculated as

$$E(J_a) = \pi J_s(M_1) + (1-\pi)\{J(M_1, M_2, T^*) + e^{-rT_*} J_s(M_2)\} \tag{4}$$

In more complete, specific parametric form $E(J_a)$ is

$$E(J_a) = \pi[(\alpha M_1)^{\delta(1-\gamma)}/r(1-\gamma)] + (1-\pi)e^{-rT_*}[(\alpha M_2)^{\delta(1-\gamma)}/r(1-\gamma)]$$
$$[1 + (r/(A-\alpha))(e^{A-\alpha)T_*} - 1)] \tag{5}$$

To find the optimal F_0 under this full treatment of uncertainty (FT), the above equation for $E(J_a)$ is used in conjunction with 3(10), the latter dealing with the simple case.

3.5   NUMERICAL CALCULATIONS

We have presented two alternatives to dealing with uncertainty and choosing the appropriate emissions level. The results show so far that emissions are higher if the "expected case" (EC) is used.[2] However, to determine the magnitude of the differences we have to resort to numerical examples, and we turn to this next. We did some limited calculations for specific examples presented in Tables 1 and 2.

In the first three columns of each table we list the parameters used in each run or set of calculations. In the fourth column we list the percentage increase in initial emissions when we change from the FTU case to the EC case. In the column EC appears the present optimal fossil fuel use when the expected value of M_c is assumed certain. In the final column FTU is the value of $F(0)$ when uncertainty is fully treated.

In all calculations we assume $M_0 = 1$, $\pi = 0.5$, production equals F^δ for M less than the critical level of CO_2.

In Table 1 the uncertain levels are $M_1 = 2$ and $M_2 = 4$. At this level of uncertainty the estimates of the proper emission level are close together. In Table 2 a much wider range of uncertainty is considered: $M_1 = 2$ and $M_2 = 8$. Therefore the differences in initial emission rates are more significant.

The emissions when an EC critical level of carbon dioxide is assumed are on the average over 30 per cent higher than in the case of FTU.

These examples add to the claim that an appropriate treatment of uncertainty is important in modelling carbon dioxide.

Table 1*: Comparison of EC and FTU with Low Variance					
Basis: $M_0 = 1$, $M_1 = 2$, $M_2 = 4$, $\pi = 0.5$					
$\delta(1-\lambda)$	r	α	% change in F_0	EC F_0	FTU
0.5	0.02	0.001	12.2	0.087	0.077
0.75	0.04	0.001	7.2	0.332	0.309
0.95	0.06	0.001	0.4	2.420	2.410
0.5	0.02	0.002	5.5	0.094	0.089
0.75	0.04	0.002	4.8	0.342	0.326
0.9	0.06	0.002	1.1	2.486	2.460

				EC	FTU
Table 2*: Comparison of EC and FTU with High Variance					
Basis: $M_0 = 1$, $M_1 = 2$, $M_2 = 8$, $\pi = 0.5$					
$\delta(1-\lambda)$	r	α	% change in F_0	EC $\quad$ FTU F_0	
0.5	0.02	0.001	46.4	0.215	0.134
0.75	0.04	0.001	34.2	0.825	0.584
0.95	0.06	0.001	24.3	6.105	4.780
0.5	0.02	0.002	39.6	0.230	0.154
0.75	0.04	0.002	32.1	0.850	0.615
0.95	0.06	0.002	24.0	6.210	4.880

* Notes: 1. Production equals F^δ for M less than critical level of CO_2.

2. Column 4 is the percentage difference in the F_0 base case. It shows the impact of treating uncertainty correctly, FTU, against the EC.

3.6   IMPACT OF UNCERTAINTY TREATMENT AND MODEL CHANGES

In Sections 4 and 5 we focused on changes in parameters. In this example we focus on two different models of the impacts of carbon dioxide, the step model versus the modified Cobb-Douglas model. In the discrete-type step model the impacts of CO_2 manifest themselves suddenly, in contrast, a modified Cobb-Douglas model shows the negative effects (of CO_2 accumulation) occurring gradually. We use the same specification of the utility function as in Section 3. Again, the model assumes that a critical level of CO_2 accumulation, M_c, exists, at which production drops to zero. M_c is used to define two variables, F_s and M_s: $M_s = M_c$- M and $F_s = F - \alpha M_c$. M_s gives the distance to the critical level of CO_2 concentration, and F_s is the distance from the level of emissions which maintains the critical CO_2 level. Consumption increases with increases in either of these variables. The production/consumption function is:

$$C = F_s^\delta \, M_s^1 \tag{1}$$

We refer to this model as a modified Cobb-Douglas model because equation (1) is a Cobb-Douglas form. From (1) it is clear that it will never be optimal to exceed the critical level of atmospheric CO_2 and that the CO_2 level will always be increasing because F will be greater than αM_c. An assumption inherent in (1) is that the economy produces less and less over time. In Figure 1 the two types of model impacts are compared.

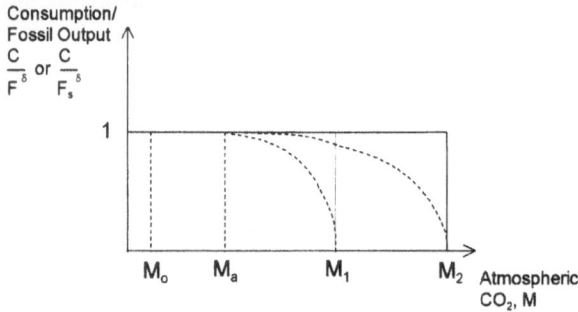

Figure 1: Alternative Models of CO_2 Impacts

94

As shown in Figure 1, in both models the current level of atmospheric CO_2 is M_o, and CO_2 currently has no impact on productivity. Further we assume that CO_2 will have no impacts until a level M_a is reached and that when this level is reached we will gain complete knowledge of all future CO_2 exist. At M_a we learn which of these is the true critical level of CO_2. Under the step model atmospheric CO_2 has no impact until M_1 or M_2 is reached. Under the modified Cobb-Douglas model, productivity is assumed to begin to decrease at M_a and to fall gradually as the critical level, M_1 or M_2 is approached. In both cases, 3(5) is maximized to determine the time at which M_a should be reached and what level of present fossil fuel use is optimal.

When the step model is assumed, we use J defined by 3(4). T_i defined by $F(t_i)=\alpha M_c e^{A(T_i-t_i)}=F(o)e^{-At_i}$ and J_s defined by 4(2) to determine the utility collected after M_a is reached. $E(J_a)$ is the expectation over $J(M_a, M_1, T_1)$ plus $J_s(M_1)$ and $J(M_a, M_2, T_2)$ plus $J_s(M_2)$ where all values are appropriately discounted.

When we assume the modified Cobb-Douglas model, after M_a is reached, the consumption path can be determined in a standard way, as in Chapter 2. However, we still need to determine the utility gathered when the Cobb-Douglas model is used. Moving from an atmospheric CO_2 level M_o towards a critical level, M_c, the utility gathered is $J_c(M_o, M_c)$ defined by:

$$J_c = \delta^{\delta(1-\gamma)} \left[\frac{r+\alpha(1-\gamma)}{1-\delta(1-\gamma)} \right]^{-(1-\delta(1-\gamma))} \left[\frac{\epsilon(M_c-M_o)^{(1-\gamma)}}{(1-\gamma)} \right] \tag{2}$$

The parameter ϵ adjusts productivity in the Cobb-Douglas model so that the step and the modified Cobb-Douglas models are comparable at M_a. $E(J_a)$ in the EC case is found using these equations and assuming that M_c equals πM_1 plus $(1-\pi)M_2$.

$E(J_a)$ is the expectation over $J_c(M_a, M_1)$ and $J_c(M_a, M_2)$ in the full treatment of uncertainty.

In tables 3 and 4 the optimal level of present fossil fuel use is calculated using both the step and the modified Cobb-Douglas model and using the expected value assumption, the EC case, and the full treatment of uncertainty, FTu. Table 3 summarizes the results, and Table 4 provides detail about the calculations.

In the first part of both tables, we chose the parameters in the middle of their expected ranges. In the second part of the tables, we chose the parameters specifically to emphasize the difference in the EC case and the FTu case. In the second part of the table, the consumption elasticity of utility is increased, which increases the importance of any differences in consumption, and the discount rate is lowered, which makes consumption in the future more important. In the first two columns of Table 3 we list the two possible critical levels. Columns (3) and (4), the most critical information is listed, the percent increase in the present use of fossil fuel when the EC case rather than the FTu case is considered. This is listed for both the step model, column (3), and the modified Cobb-Douglas model, column (4).

Table 3: Impact of Uncertainty and CO_2 Model, Summary

$M_o = 1$, $M_a = 2$, $\pi = 0.5$

Midrange parameters:

$\delta = 0.85$, $r = 0.04$, $\gamma = 0.1$, $\alpha = 0.0015$

Case	(1) M_1	(2) M_2	(3) Step % Change	(4) C-D % Change
1)	3	5	1.6	1.1
2)	3	8	5.3	2.6
3)	4	6	1.2	0.5
4)	4	8	2.5	1.2

Parameters for emphasis:

$\delta = 0.75$, $r = 0.02$, $\gamma = 0.333$, $\alpha = 0.001$

Case	(1) M_1	(2) M_2	(3) Step % Change	(4) C-D % Change
5)	2.5	7.5	16.1	14.4
6)	3	8	11.0	9.7

Notes:
1. Production equals F^δ for M less than the critical level of CO_2.
2. Column (3) is the % difference between F_o without and with the full treatment of uncertainty when the step model of CO_2 impacts is used. Column (4) is the % difference between F_o without and with the full treatment of uncertainty when the Cobb-Douglas model of CO_2 impacts is used.

In the first three columns of Table 4 we list the two possible critical levels and the expected critical level. In the next two columns, columns (4) and (5) the percent increase in the present use of fossil fuel when the EC case rather than the FTu case is considered. This is listed for both the step model, column (4), and the modified Cobb-Douglas model, column (5).

Table 4: Impact of Uncertainty and CO_2 Model									

$M_o = 1$, $M_a = 2$, $\pi = 0.5$

$\delta = 0.85$, $r = 0.04$, $\gamma = 0.1$, $\alpha = 0.0015$

(1)	(2)	(3)	(4)	(5)		(6)	(7)	(8)	(9)
M_1	M_2	$\pi M_1 +$ $(1-\pi)M_2$	Step % Change	C-D % Change		Step		Cobb-Douglas	
						EC	FTu	EC	FTu
3	5	4	1.6	1.1	$E(J_a)$	3.20	3.14	2.51	2.48
					$F(0)$	0.558	0.549	0.452	0.447
3	8	5.5	5.3	2.6	$E(J_a)$	4.81	4.57	3.85	3.75
					$F(0)$	0.835	0.792	0.666	0.649
4	6	5	1.2	0.5	$E(J_a)$	4.30	4.25	3.42	3.40
					$F(0)$	0.744	0.735	0.594	0.591
4	8	6	2.5	1.2	$E(J_a)$	5.31	5.19	4.26	4.21
					$F(0)$	0.928	0.905	0.737	0.728

$\delta = 0.75$, $r = 0.02$, $\gamma = 0.333$, $\alpha = 0.001$

(1)	(2)	(3)	(4)	(5)		(6)	(7)	(8)	(9)
M_1	M_2	$\pi M_1 +$ $(1-\pi)M_2$	Step % Change	C-D % Change		Step		Cobb-Douglas	
						EC	FTu	EC	FTu
2.5	7.5	5	16.1	14.4	$E(J_a)$	14.50	13.08	11.07	9.964
					$F(0)$	0.188	0.160	0.126	0.109
3	8	5.5	11.0	9.7	$E(J_a)$	15.58	14.53	11.95	11.15
					$F(0)$	0.211	0.189	0.140	0.127

Notes:
1. Production equals F^δ for M less than the critical level of CO_2.
2. Column (4) is the % difference between F_o in columns (6) and (7), and column (5) is the % difference between F_o in columns (8) and (9). Columns (4) and (5) show, for the step model and the Cobb-Douglas model respectively, the impact of treating uncertainty correctly, FTu, or, EC case.

In the next four columns, we list the raw data from which we derive these ratios.

This includes the expected value to be gathered after the CO_2 level M_a is reached and uncertainty is resolved, $E(J_a)$, and the initial use of fossil fuels, F_0. These values are listed for both the EC case and the FTu case and for both the step and the modified Cobb-Douglas model.

Looking at the top part of table 3 for the midrange parameter levels (cases 1 through 4), the impact of the treatment of uncertainty is not great. In this portion of the table the largest change in fossil fuel use due to a change in the uncertainty treatment is 5.3 percent, case 2. With the midrange parameters, the treatment of uncertainty is about half as important under the modified Cobb-Douglas model, as measured by the percent change in fossil fuel use (comparing columns 3 and 4 in the upper portion of Table 3).

If the parameter values are chosen to emphasize the importance of the uncertainty treatment as in the bottom of Table 3 (cases 5 and 6), two effects occur. First, of course, uncertainty has a bigger impact. Second, the impact of the step and the modified Cobb-Douglas models are more similar (compare columns 3 and 4 in the bottom portion of the table). In the more moderate examples, the influence of uncertainty was almost twice as great in the step model as in the Cobb-Douglas model. In the examples with more severe uncertainty impacts, the effects are only about 11 to 14 percent greater in the step model than in the Cobb-Douglas model. Examining 3(10) this suggest that the ratio of $E(J_a)$ in the EC case to $E(J_a)$ in the FTu case is growing more similar for the step and Cobb-Douglas models. However, an examination of this ratio has not revealed an obvious reason for the increasing similarity.

3.7 CONCLUSIONS

Uncertainty considerably complicates economic modelling, therefore, it is important to assess the value of intrinsic treatments of uncertainty. We have shown that the use of expected values of critical parameters rather than the full treatment of uncertainty can lead to a significant overstatement of the optimal present fossil fuel use. The determinations of optimal fossil fuel use are similar when there is an uncertain, critical atmospheric carbon dioxide level and when the fossil fuel resource is uncertain. In fact, when there is zero absorption of atmospheric carbon dioxide, the problems are identical. When absorption is greater than zero, the results of research on uncertain, limited, natural resources are often applicable to the carbon dioxide problem. It seems reasonable that the results can be extended to related areas of economic modelling where optimal decision making with uncertainty is called for.

REFERENCES

Conrad, J. M., 'Stopping Rules and the Control of Stock Pollutants', in Brekke, K. A. (ed.), Seminar on Uncertainty in Management of Natural Resources and the Environment, Central Statistical Bureau, Research Dep., Oslo, Norway, 1992.

Chichilnisky, G. and G. Heal, 'Global Environmental Risks', *Journal of Economic Perspectives*, 7, 1993, pp. 65-86.

Gilbert, R. J., 'Optimal Depletion of an Uncertain Stock', *Review of Economic Studies* 46, 1979, pp. 47-57.

Holly, S., and A. H. Hallet, *Optimal Control, Expectations and Uncertainty*, Cambridge University Press: Cambridge, 1989.

Henrion, M., and M. G. Morgan, Uncertainty: *Guide to Dealing with Uncertainty in Quantification Risk and Policy Analysis*, Cambridge University Press: Cambridge, 1990.

Ingham A., and A. Ulph, 'The Economics of Global Warming', in Bennett, J. and W. Block (eds.), *Re-approaching Economics and the Environment*, Australian Institute for Public Policy: Canberra, pp. 223-248, 1991.

Kolstad, C. D., 'Looking vs. Leaping. The Timing of CO_2 Control in the face of Uncertainty and Learning', *International Workshop on Costs, Impacts and Possible Benefits of CO_2 Mitigation*, IIASA, Laxenburg, Austria, 1992.

Malliaris, A. G. and W. A. Brock, Stochastic Methods in Economics and Finance, North Holland: Amsterdam, 1982.

Peck, St. C. and Th. J. Teisberg, 'Global Warming Uncertainty and the Value of Information: An Analysis using CETA; *Resource and Energy Economics*, 15, 1993, pp. 71-97.

Footnotes

[1] Using the necessary conditions, F(f), as a function of αM_c is found:

$$f(t) = \alpha M_c e^{A(T-t)} = F(0)e^{-At} \tag{A1}$$

Relating to pertinent results of Chapter 2, the time at which M equals M_c, T is defined by:

$$M_c = \int_0^T e^{-\alpha(T-t)} \phi (q(t))dt + e^{-\alpha T}M_0 \tag{A2}$$

Equation (A1) for F(t) can be substituted into (A2) and a simple expression in T derived:

$$\left[\alpha M_c/(A-\alpha) M_0\right] \left[e^{(A-\alpha)T} - 1\right] + e^{-\alpha T}(M_c/M_0) = 0 \tag{A3}$$

T^* can be determined on the basis of (A3).

[2] The theorems developed by Gilbert (1979) assure that emissions are higher if the expected case is used. However, the theorems do not suggest the magnitude of the differences. Interestingly enough, there are related results by Conrad (1992). Using Ito calculus, Malliaris and Brock (1982), Chapter 2, he shows that the optimal stock of a non-degradable pollutant subject to stochastic environmental cost will be less than the optimal stock for a non-degradable pollutant with known damage.

CHAPTER 4

LONG-RUN INVESTMENT AND ENDOGENOUS TECHNICAL PROGRESS:
DYNAMIC AND VINTAGE-TYPE MODELS

4.1 INTRODUCTION

In this chapter we set out to model the economic aspects of the CO_2 problem under endogenous technical progress (Romer, 1990). Such models appear more natural and provide increased flexibility and realism for policy-making purposes.

We begin with a comparison of the models with those developed in Chapter 2. The analysis considers possible long-run equilibrium solutions and possible approach paths to equilibrium.

We present a solution for a constant, long-run growth path of the model and an examination of the effects of parameters on the ratio of "knowledge" to capital and on the growth rate along this path.

The model introduced here bears a strong resemblance to the simple model with exogenous technical progress in Chapter 2; however, here we specify the manner in which technical progress occurs. In this model the economy does not acquire technical progress for free but must invest in knowledge and physical capital. There is an additional economic rationale for including capital accumulation and knowledge in the analysis of the CO_2 problem, that is, a compensation argument for intertemporal or intergenerational equity (Spash and d'Arge, 1989). If a given fossil fuel consumption and possible ensuing greenhouse warming cannot be avoided in this generation or the next, then at least part of the capital and knowledge should be put to mitigate the effects, or to create technologies which future generations could use to protect themselves against any harmful effects. This is part of an insurance policy on CO_2 strategies (Manne and Richels, 1992a; Schelling, 1991).

This model is more flexible because the level of investment is not fixed in advance but is determined within the optimization process. A limitation of this model is that only neutral technical progress is allowed, but its usefulness can be defended on the aggregate level, as intended (V. K. Smith, 1974, Ch. 2). The three control variables in the model are the level of fossil fuel use, F; the level of investment, I, and the distribution of investment, θ. Atmospheric CO_2, M; knowledge, S, and capital, K, are all state variables. F, M, S and K all determine production, Y.

As in previous models the objective is to maximize utility discounted at the social discount rate, r, over an infinite horizon.

We assume that the utility function has a constant consumption elasticity of utility with elasticity γ as shown below.

$$U(C) = \begin{matrix} (C^{1-\gamma})(1-\gamma), \gamma \neq 1 \\ lnC, \gamma = 1 \end{matrix}$$

(1)

The effects of CO_2 and energy use are determined by the function h(F,M) and are separable from the effects of knowledge and capital. A Cobb-Douglas type term determines the impact of knowledge and capital on production. In this term the elasticities of production with respect to knowledge and capital are μ and v, respectively. Production is:

$$Y = h(F,M) S^\mu K^v$$

(2)

Because of the possibility of investment, this model is significantly different from those presented before. Consumption is C = Y - I, and investment is divided between knowledge and capital. θI is invested in knowledge and $(1-\theta)I$ is invested in capital. We assume that the control variables F and I are both greater than or equal to zero. θ is limited between zero and one which means that knowledge cannot be changed to capital, capital cannot be changed to knowledge and neither knowledge nor capital can be consumed. The atmospheric CO_2 level changes as in previous models,

$$dM/dt = F - \alpha M$$

Capital and knowledge differ in two major ways. First, capital depreciates at the rate ρ while knowledge does not depreciate. Second, the change in capital is a linear function of capital investment, but the change in knowledge is a non-linear function of knowledge investment. These assumptions are specified by

$$dS/dt = (\theta I)^\sigma$$

(3)

where σ can be seen as a "society-dependent" transformation parameter of human capital investment into knowledge.

$$dK/dt = (1-\theta)I - \rho K$$

(4)

For reasons of convenience, we refer to the derivative of the right-hand side of (3) with respect to I divided by θ as the *effective knowledge investment*, ESI. ESI is the marginal change in dS/dt per unit of investment in knowledge, that is:

$$ESI = \sigma(\theta I)^{\sigma - 1}$$

$$(5)$$

4.2 FIRST-ORDER NECESSARY CONDITIONS

The Hamiltonian, H, for this problem is

$$H = U(C) - q(f - \alpha M) + \zeta(\theta I)^2 + \kappa((1-\theta)I - \rho K \tag{1}$$

Using optimal control we derive the necessary conditions in integral form. The first three equations determine the values of the adjoint variables or the shadow prices of fossil fuels, knowledge and capital. The shadow price of fossil fuels, q, is

$$q = -\int_t^\infty e^{-(r+\alpha)(\tau-t)}$$
$$C^{-\gamma}(Y/h(F,M))(\delta h/\delta M)\, d\tau \tag{2}$$

As in earlier models the shadow price of energy at time t, q(t), represents the total future loss due to a one unit increase in atmospheric CO_2. The marginal disutility of increased CO_2 is the marginal utility of consumption times the marginal decrease in production with an increase in atmospheric CO_2. Equation (1) shows that the total future loss is the marginal disutility of increased CO_2 discounted at the social discount rate plus the absorption rate of CO_2 and summed over time.

$$\zeta = \int_t^\infty e^{-r(\tau-t)} C^{-\gamma}\mu(\eta/s)\, d\tau \tag{3}$$

$$\kappa = \int_t^\infty e^{-(r+\rho)(\tau-t)} C^{-\gamma}v(Y/K)\, d\tau \tag{4}$$

ζ and κ represent the long-run values of a unit increase in knowledge and capital, respectively. ζ is the value of a unit increase in knowledge as its marginal utility discounted at the social discount rate and summed over time. In a similar way, we define κ as the value of a unit increase in capital, discounted at the social discount plus the depreciation rate. The marginal utility times the marginal productivity of energy must be less than or equal to the shadow price of CO_2, q

$$C^{-\gamma}(Y/h(F,M))(\delta h/\delta F) \le q \tag{5}$$

105

with a strict inequality holding if F=0. Similar relations hold between the adjoint variables ζ, κ and the marginal utility of consumption. The reasoning goes like this. Because capital changes as a linear function of investment, the value of capital also measures the value of capital investment. However, the value of investment in knowledge is measured by the effective knowledge investment times the value of an increase in knowledge. Thus we can state the sum of the values of investment in knowledge and capital weighted by the distribution of investment in each must be less than or equal to the marginal utility of consumption.

$$\theta\,(ESI)\,\zeta + (1-\theta)\,\kappa \leq C^{-\gamma}$$

$$(6)$$

If the 'less than' condition holds, I equals 0. Furthermore, if σ were less than one and I were equal to 0 the effective knowledge investment would be infinite and (6) would be violated. Therefore, if σ is less than one then I is always greater than 0,

$$
\begin{aligned}
(ESI)\,\zeta &> \kappa, & \theta &= 1 \\
(ESI)\,\zeta &= \kappa, & 0 &\leq \theta \leq 1 \\
(ESI)\,\zeta &< \kappa, & \theta &= 0
\end{aligned}
$$

$$(7)$$

(7) assures that we invest in both knowledge and capital only when it is equally effective in each. If the value of investment is unequal in knowledge and capital, we invest only in the more valuable factor. Adjoint variables must be continuous except at the boundary values of the associated state variables. This implies that C is continuous, and it follows that Y, I, M, K, S and F are continuous.

4.3 LONG-RUN EQUILIBRIUM

We first establish the existence and characteristics of two possible long-run equilibria. In one equilibrium, consumption growth is zero, in the second equilibrium, the economy grows exponentially. In equilibrium the energy-CO_2 sector of the economy behaves as in the simple static model with given exogenous technical process in Chapter 2.

Even if the economy grows, F and M are constant in the equilibrium. The necessary conditions can be considerably simplified. We first examine the conditions on the energy CO_2 sector. If F is greater than 0, then from 2(2) and 2(5):

$$q^* = [r + \alpha + (\delta h / \delta M)] / (\delta h / \delta F)$$

(1)

As previously, rates are designated by *. If $\theta < 1$, from 2(7), (ESI)$\zeta \le \kappa$. We can easily solve 2(6) for κ:

$$\kappa = C^{-\gamma} \ for \ \theta < 1$$

(2)

Let g denote the rate of consumption growth. By differentiating (2) with respect to time combining (2) with 2(4) and restating the equation in terms of rates of change

$$-\gamma g = r + \rho - \nu (Y/K) \ for \ \theta < 1$$

(3)

If $\theta > 0$, then

$$\zeta = C^{-\gamma} / ESI \ for \ \theta > 0$$

(4)

and

$$-\gamma g (1-\sigma) \ (I^* + \theta^*) = r - \mu (ESI) \ (Y/S) \ for \ \theta < 0$$

(5)

The system of necessary conditions has been considerably simplified. Both ζ and κ have now been solved out of the system. In our proposed equilibria, g equals zero or is constant and greater than zero; C, Y, and I all grow at g, and θ is constant.

The equal growth rates of C,Y, and I allow equation 1(3) to be satisfied. Equations 1(4), 2(5) and (1) can be satisfied if q increases at a constant rate of g (1-γ) and F and M are constant. As noted earlier, we assume that h has the same properties as f in the simple model. It follows that the pair of equations 1(4) and (1), can be solved in the same manner as the similar equations for the simple model with exogenous exponential technical progress. F, M and h(F,M) are constant, and the following relation holds:

$$M = \alpha F$$

(6)

Along a line in the F-M plane where h is constant

$$dM/dE = 1/(r+\alpha-g(1-\gamma))$$

(7)

The familiar equations in (6) and (7) are very important, the intersection of the lines defined by these equations determines the equilibrium values of F and M in all cases.

If no investment is made in the capital, the ratio of knowledge to capital will grow as capital stock shrinks, consequently the marginal product of capital will grow; and when the marginal product of capital is large enough, it will always be optimal to sacrifice some consumption for investment in capital. In equilibrium, investment must be greater than 0 and θ must be greater than 0.

We can now define the static equilibrium. In the static equilibrium g equals zero, and investment is made in capital only. I equals ρK. The ratio of S^μ to K^ν is treated as a single variable and (3), (6), (7) form a system of three equations in three unknowns which can be solved for the equilibrium.

We now examine an equilibrium in which g > 0. By our previous argument, investment in capital must be greater then zero. By a similar argument, if the economy and capital stock are growing, investment must occur in knowledge or the marginal product of knowledge goes to infinity. A growth equilibrium is, therefore, also a balanced growth equilibrium in which knowledge and capital both receive investment and both (3) and (5) hold. Next we look back at 1(5) and 1(6). If θ is constant, S increases at a rate of gσ and K increases at a rate of g. Examining the production function 1(2) we reach the conclusion that a balanced growth path equilibrium exists only if a critical condition is met.

108

Assume $\sigma\mu+\nu = 1$

$$\tag{8}$$

This assumption might be more easily understood by comparing the effects of a constant rate of investment in economies where (8) does and does not hold. Suppose that current inputs and the rate of investment are constant in each of the following cases:

(i)   If $\sigma\mu+\nu$ equals one, the economy grows uniformly at the investment rate. Most importantly, the marginal products of knowledge and capital and the fraction of production invested are constant.

(ii)   If $\sigma\mu+\nu$ is greater than one, the marginal products of knowledge and capital continually rise and the fraction of production invested continually falls.

(iii)   If $\sigma\mu+\nu$ is less than one, the opposite effects would occur.

A better consideration of the reasonableness of the assumption would require data on historical trends in output, current input productivity and use, and knowledge and capital investment. For example, if such data showed that historically production, research and capital investment all grew at a common rate but that the productivity of current inputs stagnated, the assumption that $\sigma\mu+\nu$ equals one would be supported. On the other hand, if production, research, capital investment, and the productivity of current inputs all grew at a common rate, this would suggest that $\sigma\mu+\nu$ is less than one. After this digression, using the relation in (8) we can state

$$Y/K = h(F,M) \ [S/K^\sigma]^\mu$$

$$\tag{9}$$

and

$$(ESI) \ (Y/S) = h(F,M) \ \sigma(\mu\sigma g)^{(\sigma-1)/\sigma} \ (K^\sigma/S)^{\mu/\sigma}$$

$$\tag{10}$$

(3) and (5) can be rewritten as:

109

$$-\gamma g = r + \rho + \gamma h(F, M) \, (S/K^{\sigma})^{\mu}$$

$$(11)$$

$$-\gamma g = r + (1-\sigma) g - \mu\sigma(\sigma g)^{(\sigma-1)/\sigma} \, h(F, M) \, (K/S)^{\nu/\sigma}$$

$$(12)$$

Equations (11) and (12) derived from (3) and (5) have straight-forward interpretations.

In both we have one term giving the marginal value of investment. For capital, this equals the marginal product of capital; and for knowledge, this equals the marginal product of knowledge times the marginal effectiveness of investment. The investment in both cases must have an equal return to the drop over time in output value and production. Output value falls due to discounting and the change in marginal utility with growth in consumption. Production from a unit of capital disappears due to depreciation. Equations (6), (7), (11), (12) are a system of four equations in four unknowns M, F, g, and the ratio of K^{σ} to S. These equations define the growth equilibrium, in order to lend it to policy directed computational support we have to further specify and simplify the model. Such growth equilibrium paths essentially depend on the Cobb-Douglas technology (Dixit, 1977, 1990).

4.4 SIMPLIFICATION OF MODEL FOR COMPUTATIONAL PURPOSES

A simple model allows us to examine the two equilibria in more detail and determine which of the equilibria is appropriate.

Again we use the step model of CO_2 impacts. We assume that a critical CO_2 level exists, M_c, below which atmospheric CO_2 has no effect upon production and above which production drops to zero. Further we assume that there is a fossil fuel productivity factor (ϵ) constant at δ and the fossil fuel sector is multiplied by a scaling factor λ (λ is only introduced for the computational analysis, it is of no importance to the economic analysis). In this model the equilibrium use of energy is αM_c.

To simplify notation, the variable ω is equal to the energy sector output in equilibrium, defined as $\lambda (\alpha M_c)^\delta$. We also assume that, as with capital, knowledge increases approximately linearly with investment or research; therefore, σ equals one. This follows from the assumption made in 3(8) that $\nu = 1-\mu$. These assumptions are expressed as

$$Y = \lambda F^\delta S^\mu K^{1-\mu}, \quad M \leq M_c$$
$$0 \qquad\qquad, \quad M > M_c$$

$$(1)$$

$$\frac{dS}{dt} = \theta I$$

$$(2)$$

With these simplifications 3(11) and 3(12) can be restated as

$$\lambda F^\delta (1-\mu) [S/K]^\mu = r + \rho + \gamma g$$

$$(3)$$

$$\lambda F^\delta \mu [K/S]^{1-\mu} = r + \gamma g$$

$$(4)$$

4.5 CLASSIFICATION OF EQUILIBRIA

4.5.1 Stationary Equilibria

In stationary equilibrium only capital receives investment, $\theta = 0$, the CO_2 level equals M_c, and investment equals K. 3(6) and 4(3) form a two-equation system in two unknowns, F and K/S. The equilibrium is similar to that in many growth models in that the ratio of the stocks determines the equilibrium. F equals αM_c. The ratio of K/S is determined by the equation 4(3). The specific level of knowledge, capital and consumption at the equilibrium is determined by the initial point.

A comparative statics analysis of equilibrium can be made. The equilibrium level of fossil fuel sector output, ω, increases with increases in α, M_c, and δ. These parameters either make fossil fuels more productive or CO_2 limits less severe. When g equals zero from 4(3), four parameters, ω, μ, ρ and r affect the equilibrium value of K/S. Because greater productivity reduces the importance of capital depreciation, increases in ω increase the marginal product of capital at any given ratio of capital to knowledge. This causes an increase in the level of the capital relative to knowledge. Conversely, increases in the social discount rate, r, increase the value of current consumption versus investment in capital and increase the shadow cost of capital. Increases in depreciation, ρ, cause capital to disappear more rapidly and likewise increase the shadow cost of capital. Increases in r or ρ result in a lower relative use of capital. The impacts of changes in μ are uncertain.

4.5.2 Balanced Growth Equilibrium

In the balanced growth equilibrium F again equals αM. Both 4(3) and 4(4) hold and these two equations can be solved for the ratio of K to S and for the rate of economic growth, g. Production Y equals $\omega S^{\mu} K^{1-\mu}$. Using 1(3) and 1(4), I and ω can be found.

By taking the total differential of 4(3) and 4(4) we can examine changes in the equilibrium values of g and the ratio of K to S with respect to changes in parameter values. Again the impact of μ on either value is uncertain. An increase in the productivity of the energy sector as measured by ω causes both the rate of economic growth, g, and the ratio of capital to knowledge to increase. Again, a more productive economy makes depreciation less important. An increase in consumption elasticity of utility, γ, causes the economy to grow more slowly but has no effect on the capital to knowledge ratio. An increase in the social discount rate has the same effects as an increase in γ. An increase in the capital

112

depreciation rate lowers the equilibrium growth rate and the capital to knowledge ratio.

4.5.3 Optimality of Equilibrium

We assume that optimal behaviour leads to an equilibrium. However, the proofs of stability and optimality, used before, cannot be applied to this problem, because the Hamiltonian is not differentiable with respect to atmospheric CO_2 at the equilibrium. We know that for optimality the fossil fuel use and the atmospheric CO_2 level must converge to αM_c and M_c, respectively.

If a path does not raise CO_2 to the critical level, it leaves unused capacity in the economy. For any such path we can construct a dominant path which does raise CO_2 to the critical level. If the CO_2 level remains at the critical level, the problem is reduced to a capital two-sector growth model (Uzawa, 1988) where it will be optimal for knowledge and capital to converge to an equilibrium. If it is not optimal for CO_2 to remain at the critical level then oscillations down from and back to the critical level must be optimal. Such oscillations seem highly unlikely to be optimal, because of both discounting and the curvature of the utility function. Oscillations cause periods of reduced consumption, discounting tends to move higher levels of consumption to earlier time periods. The greater the curvature of the utility function the greater the loss in marginal value as consumption moves from low points to high points. In general, the curvature of the utility function tends to smooth out the consumption stream and eliminate cycles in consumption.

Within our means we were unable to prove that cycles are not optimal though tools along this line could be put to test (Goodwin, 1990). Instead, in what follows, I will use a simple graphical representation to argue that at least one of the equilibria classified represents long-run optimal behaviour if an optimum exists.

If the economy is very productive, an optimum defined by regular maximisation does not exist. If the utility growth rate, $g(1- \gamma)$, is greater than the social discount rate, r, welfare will be infinite. In this case the investment required to maintain the growth rate will be greater than the production.

Figures 1 and 2 illustrate how the appropriate equilibrium is determined. The figures show levels of knowledge, OA, and of capital, OB, both as lines in the knowledge-capital plane. Along the knowledge line the marginal product of knowledge (MPS) equals the social discount rate, r, and along the capital line the marginal product of capital (MPK) equals the sum of the social discount rate, r, plus the depreciation rate, ρ

$$MPK = \lambda \ F^{\delta} \ (S/K)^{\mu}, \qquad MPS = \lambda \ F^{\delta} \ (K/S)^{1-\mu}$$

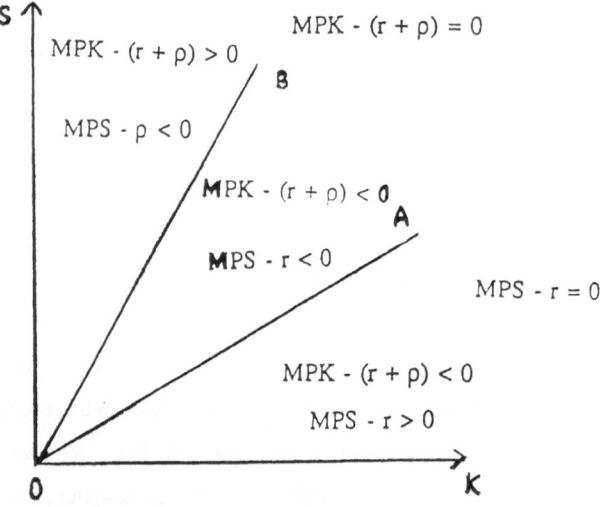

Figure 1: Stationary Equilibrium

In Fig. 1 the capital line lies above the knowledge line. If MPK is greater than or equal to r+ρ, MPS must be less than r. In this situation g along the equilibrium path is negative. Since it is never optimal for knowledge to decrease this is clearly not an optimal equilibrium.

$$\text{MPK} = \lambda \ F^{\delta} \ (S/K)^{\mu}, \qquad \text{MPS} = \lambda \ F^{\delta} \ (K/S)^{1-\mu}$$

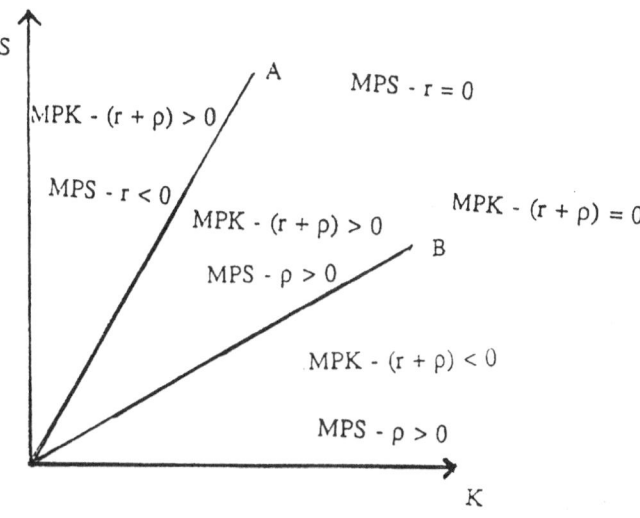

Figure 2:   Possible Growth Equilibrium

In Fig. 2 the knowledge line, OA, lies above the capital line, OB, and a line exists along which both the marginal product of knowledge minus the social discount rate and the marginal product of capital minus the sum of the social discount rate plus the depreciation rate are equal and greater than zero. In this case g is positive and the balanced growth equilibrium is optimal.

4.6 CHARACTERIZATION OF EQUILIBRIUM

We now analytically characterize the rates of change of energy use, knowledge, and capital as the equilibria are reached. The solution of these equations requires the continuity of consumption and investment.

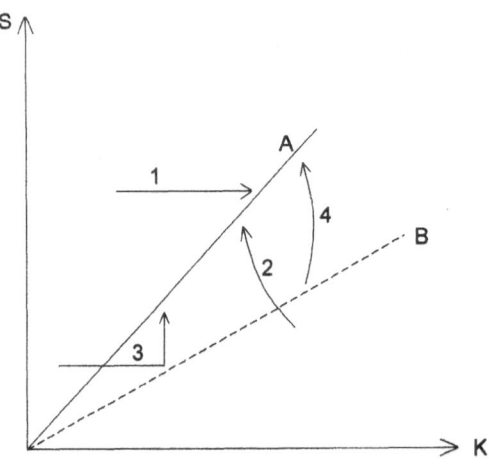

Figure 3: Transition Paths to Optimal Growth

Fig. 3 illustrates possible transition paths to a balanced growth equilibrium, labelled 1 to 4. When fossil fuel use equals α M_c, OA in the figure is the line along which balanced investment takes place. When the level of output from the fossil fuel sector is greater than the equilibrium output, OB is the line along which balanced investment will proceed.

As indicated by the figure the line OB lies below and moves up to coincide with OA as fossil fuel use moves toward equilibrium. Above line OA the ratio of knowledge to capital is too high and investment takes place in capital only, path 1, below OB the opposite results in path 2.

The possibility exists of combined paths in which we first invest only in capital and then invest only in knowledge, path 3. Along path four the ratio of knowledge and capital,

116

by reaching OA, stays on OA as OA moves. The possibility of being on such a path will be further examined later on.

Paths to a stationary equilibrium would be similar. However, when converging to a stationary equilibrium a region exists in which it is not optimal to invest in either knowledge or capital, the region between OA and OB in Fig. 1. In this region the capital stock is allowed to shrink as the equilibrium is approached.

At the moment equilibrium is reached, the continuity of production and consumption determine investment and the continuity of the left hand sides of 4(3) and 4(4) determine g. Knowing g and I allows us to solve for the rates of change of the variables just as the equilibrium is reached and allows the analysis of this small portion of the transition path. Because of the similarity of the approach to the growth and stationary equilibrium, we will only discuss the approach to the growth equilibrium.

From 3(1) we derive a condition on fossil fuel use which holds when CO_2 is not at its critical level:

$$C^{-\gamma} \partial (Y/F) = q_o^\theta$$

(1)

By differentiating both sides of (1) and dividing out common elements, we obtain:

$$-\gamma g + \eta^* - F^* = r + \alpha$$

(2)

4(1) provides an expression for Y in terms of F, S and K, and it can be differentiated with respect to time and substituted into (2) obtaining

$$\mu S^* + (1-\mu) K^* - (1-\delta) F^* = \alpha + r + \gamma g$$

(3)

Because investment is continuous, the total investment in knowledge and capital, the sum of the right-hand sides of 1(4) and 4(2) must be equal just before and after the equilibrium is reached. We can then derive the following equation which holds as equilibrium is reached.

$$SS^* + KK^* = g(S+K)$$

(4)

117

Along regular transition paths to optimal growth, investment is *either* in knowledge *or* capital alone when equilibrium is reached. Knowledge and capital reach their equilibrium ratio after or, less likely, just as fossil fuel use reaches its equilibrium level. In all of these cases, (3) and (4) are sufficient to determine the rates of change of all variables. It is much more difficult to determine the possibility of balanced investment equilibrium. If equilibrium can be reached with investment in both knowledge and capital, we can equate 3(11) and 3(12), differentiate with respect to time, and develop

$$(\mu\rho+r+\gamma g) S^* - (\mu\rho+r+\gamma g) K^* + \delta\rho F^* = 0$$

(5)

(3) - (5) form a system of three linear equations in three unknowns. One method of solving this system is to invert the matrix of coefficients of the three variables S^*, K^* and F^* and multiply the vector of the right-hand sides of these equations. A first step in finding the inverse is to find the determinant of the matrix, that is

$$D = \delta\rho S - (S+K) (\mu\rho + (1-\delta) (r+\gamma g))$$

(6)

We then can calculate three equations to determine S^*, K^* and F^* as the equilibrium is approached.

$$S^* = g - [\delta\rho K(\alpha+r-(1-\gamma) g)] / [\delta\rho S - (S+K) (\mu\rho + (1-\delta) (r+\gamma g))]$$

(7)

$$K^* = g + [\delta\rho S(\alpha+r-(1-\gamma) g] / \delta\rho S - (S+K) (\mu\rho + (1-\delta) (r+\gamma g))]$$

(8)

$$F^* = [(K+S) (\mu\rho+r+\gamma g) (\alpha+r-(1-\gamma) g)] / [\delta\rho S - (S+K) (\mu\rho + (1-\delta) (r+\gamma g))]$$

(9)

It should be clear that if D is positive, the equilibrium cannot be reached along a balanced investment path. If D is positive, then fossil fuel use is increasing. To reach equilibrium, fossil fuel use must be greater than αM_c and increasing at equilibrium, CO_2 must either exceed the critical level or the fossil fuel path must be discontinuous. Neither of these can be optimal. D must therefore not be positive. A high social discount rate encourages high early use of fossil fuels and makes the possibility of a positive D less likely. High values of the consumption elasticity of utility encourage high early fossil fuel use to level out consumption in a growing economy and affect D in the same manner as the social discount rate. The role of the fossil fuel productivity factor, δ, and depreciation, ρ, are less clear. (5) shows that if δ and ρ are high the ratio of knowledge to capital must change rapidly along the balanced growth path. This suggests that, when fossil fuel use is dropping and δ and ρ are large, the investment required to maintain a balanced approach path is too high to be optimal. If D is zero, an infinity of solutions or no solutions may exist. Further, very small variations in parameters, while allowing us to find unique solutions to the problem, are likely to cause completely varying answers.

4.7 NUMERICAL EXAMPLES

Further study of this model requires the use of numerical examples. The model is not detailed or realistic enough to allow an estimation of parameters, therefore we examine the behaviour of the model over a wide range of parameter values. Because our purpose is to discover anomalies in the model's behaviour and to examine the shapes of possible optimal paths, the ability to scale the model is not of immediate concern. One major feature of equilibrium discussion has been the balanced growth equilibrium. In this context we see that the individual parameters of the energy sector, λ, α, M_c and δ do not affect the equilibrium. It is only necessary to specify the total output of the energy sector in equilibrium, ω. The other parameters we need to specify are the consumption elasticity of utility, γ, the discount rate, r; the elasticity of production with respect to knowledge, μ, and the depreciation rate of capital, ρ. In order to reduce the number of cases examined and because the value of the capital depreciation rate is less uncertain than the others, the depreciation rate of capital was set at 0.1 in all cases. The range of values of the parameters are listed:

```
ρ=0.1
μ=0.1-0.4,r=0.01-0.1,γ=0.5-0.95
ω=0.1-0.4
```

All combinations of the extreme values of the parameters provide sixteen cases which were examined for possible balanced equilibria. In four cases the optimum is not defined because $(1-\gamma) g$ is greater than r. These cases are characterized by high equilibrium energy sector output, ω, which results in a high growth rate. When this is coupled with either a low discount rate or low relative risk aversion, a finite optimum does not exist. In six more cases the balanced growth rate is negative, this condition is associated with low fossil fuel sector output. This result suggests that, when fossil fuel use is severely limited due to CO_2 and substitutes are not available, the economy may stagnate even if opportunities for investment exist. The cases in which the balanced growth path represents a possible optimal equilibrium are listed below:

Table 1: Cases, Growth Rates, K/S Ratios

	Cases					
Parameter	1	2	3	4	5	6
μ	0.10	0.40	0.40	0.40	0.40	0.40
r	0.01	0.01	0.01	0.01	0.10	0.10
γ	0.95	0.50	0.95	0.95	0.50	0.95
ω	0.40	0.10	0.10	0.40	0.40	0.40
Growth g(%)	10.60	1.10	0.54	14.85	10.10	5.34
K/S	1.55	0.20	0.20	0.90	0.90	0.90

The examples show a wide variation in both the equilibrium growth rate and the equilibrium capital to knowledge ratio. Cases 2 and 3 again illustrate that low energy sector output causes low equilibrium growth rates. Cases 5 and 6 show how sensitive the model is to the curvature of the utility function. When the consumption elasticity of utility, γ , nearly doubles, the optimal growth rate is cut nearly by half. If progress and growth continue in spite of CO_2, the present generation is the poorest generation. The optimality of lowering fossil fuel use and/or consumption in the present to promote future growth is tempered by our desire to raise consumption for the poorest consumers, ourselves. Therefore, a high consumption elasticity leads to higher present consumption and lower growth.

Other relationships in Table 1 are not as transparent. As shown in 4(3) and 4(4) the marginal product of knowledge minus the marginal product of capital must equal the rate of depreciation ρ , in equilibrium. Increases in μ raise the marginal product of knowledge and lower the marginal product of capital, therefore, to maintain a difference in the marginal products equal to ρ , the K/S ratio must fall when μ rises. This is reflected in cases 1 and 4. The level of energy sector output is low, the K/S ratio becomes very low to maintain the required difference in marginal products, as in cases 2 and 3. Capital depreciation is in a sense a fixed cost of maintaining capital. When the energy sector is large this fixed cost is not too important. However, when energy sector output and growth is low, the fixed cost is important and the level of capital fails to raise the marginal product of capital. This suggests a rule of thumb that if the economy slows down due to CO_2 or other influences, a shift away from inputs with substantial fixed costs would possibly occur. Furthermore, we can conclude: when the fossil fuel sector output is shrinking the K/S ratio must be changing

121

to maintain the difference in the marginal products. It follows that the rate of change in capital and knowledge will be very different in these cases.

If the fossil fuel output shrinks rapidly to low levels, the structure of the economy must change very rapidly. Non-depreciating assets such as knowledge must increase rapidly compared to regular capital assets. A fundamental shift to a service economy? The economic interpretation of the conditions on balanced investment approaches to the balanced growth equilibrium is limited because of the complex expression for the variables' rates of change. We use numerical examples to gain some insights into possible balanced approach paths. In all examples production is normalized at one. To determine the rate of change in fossil fuel use the absorption rate of carbon dioxide and the fossil fuel productivity factor must be specific. In all these examples α is 0.001 and ϵ is 0.8.

The rates of change as the equilibrium is neared are listed in Table 2.

Table 2: Rates of Change as Equilibrium is Approached, Balanced Growth							
$\alpha = 0.001$ $\epsilon = 0.8$							
	Case	1	2	3	4	5	6
Parameters:	μ	0.10	0.40	0.40	0.40	0.40	0.40
	τ	0.10	0.01	0.01	0.01	0.10	0.10
	γ	0.95	0.50	0.95	0.95	0.50	0.50
	ω	0.40	0.10	0.10	0.40	0.40	0.40
Results:	g(%)	10.62	1.01	0.53	14.76	10.05	5.29
	S*	0.28	0.01	0.00	0.15	0.17	0.19
	K*	0.08	0.03	0.04	0.14	0.02	-0.1
	F*	-0.52	0.01	0.02	-0.03	-0.35	-0.67

The most significant feature of this table is that in the three cases with low discount rates - cases two, three, and four - fossil fuel use is increasing or decreasing very slowly. The effect of the social discount rate of fossil fuel was discussed earlier. Growth in consumption along a balanced equilibrium is too slow to cause fossil fuel use, higher initial investment, and higher consumption growth. The higher fossil fuel use would cause CO_2 to reach equilibrium before knowledge and capital.

As noted earlier, when both receive investment, the marginal products of knowledge and capital must be different by the depreciation rate, δ. We also noted earlier that low energy sector output reduced the differences in these marginal products. This same relationship is evident here in cases one, five and six. When the fossil fuel sector output is shrinking the ratio of capital to knowledge must be changing to maintain the difference in the marginal products; it follows, that the rate of change in capital and knowledge will be very different in these cases. If the fossil fuel output shrinks rapidly to low levels, the structure of the economy must change very rapidly. Non-depreciating assets such as knowledge must increase rapidly compared to regular capital assets.

4.8 MORE ON ENDOGENOUS TECHNICAL PROGRESS

With a view to generalizing the class of models on optimal energy use, as presented in Chapter 2, we develop a unique model in which prices influence the pattern of technical progress and furthermore non-neutral technical progress is possible. This model has some more realistic features.

First, it is a vintage-type model; that is the fossil energy input required by a piece of capital (equipment) is determined by the year in which it is purchased and cannot be changed after the capital is in place (Solow, 1964 and 1988).

Second, research can be performed to improve the productivity of new capital or to reduce its energy requirements. The output required, the research budget, and cost of energy and capital are all exogenous. The objective is to minimize the cost of production.

The first push toward generalization involves the departure from neutrality of technical progress. The assumption of neutral technical progress is quite common. Neutral technical progress may be seen as a natural process, as a continuation of a past trend, or may be assumed for its simplicity.

If we look for economic factors as sources of explanation for the pattern of technical progress, we should take into account the different age structure of the capital stock. A machine in place usually has a limited capacity for changes in its mix of inputs.

In order to substantially change the ratio between the inputs to the production process, the capital stock must change. A vintage model of capital, in which the date of purchase determines the operating characteristics of capital, accounts for the inflexibility of capital.

In what follows, we develop two models which differ mainly in their fossil fuel requirements. In each model, a special or preferred ratio of technical coefficients exists and particular price patterns cause research to move the technology toward this ratio. We examine how this ratio changes under different price patterns. Finally, we examine a variation from these basic models which allows us to discuss CO_2 limits.

The remainder of the this chapter is organized as follows:

We first develop several models which reasonably represent the influence of economic factors on technical change. We show the sensitivity of the optimal research pattern to assumptions about the relation of energy use to output and the research budget.

Such models could well form an essential part of policy-oriented energy models in which capital stock vintages generate major changes in technology and productivity (Scherago et al, 1990, Jorgenson, 1989; Ingham et al., 1987).

4.9 A VINTAGE MODEL

A basic feature of this model is that the portion of research devoted to each input, energy and capital, is an endogenous variable. As technical progress occurs, the characteristics of machines change. Each machine is distinguished by its vintage. $T(t)$ is the age of the oldest machine or equipment in use at time t. $M(t)$ is the life of a machine bought at time t. By definition,

$$M(t) = T(t+M(t))$$
(1)

Let the scale parameters for the oldest equipment and the equipment in use be defined by:

$$dT/dt = \tau, dM/dt = \lambda$$
(2)

$Y(k)$ is the production from all machines bought at time k. $I(k)$ is the number of machines purchased at time k. $F(k)$ is the energy used by machines of vintage k. The critical energy assumption made is that energy use is proportional to the output level.

$$Y(k) = Min(A_1(k) I(k), A_2(k) F(k))$$
(3)

$A_1(k)$ and $A_2(k)$ are the Leontief type technical coefficients.

If the technology is used efficiently the conditions below hold:

$$Y(k) = \begin{cases} A_1(k) I(k), & k \geq t-T(t) \\ 0, & k < t-T(t) \end{cases}$$
(4)

$$F(k) = \begin{cases} \dfrac{Y(k)}{A_2(k)}, & k \geq t-T(t) \\ 0, & k < t-T(t) \end{cases}$$
(5)

Because, in this context, we emphasize the nature of technological progress, the production level $Y^*(t)$ and the total research budget $R^*(t)$ are assumed exogenous. The objective is to minimize all future discounted costs. The problem is described in the following equations:

$$J= \mathop{Min}_{Y,\tau,\theta} \int_{t^*}^{\infty} \exp(-r(t-t^*)) \left\{ p_1(t)\frac{Y(t)}{S_1} + p_2(t) \int_{t-T(t)}^{t} F(k)\,dk \right\} dt \tag{6}$$

Future costs have two components. Machines sell at a price of p_1 and the first term in the bracket of (6) is the expenditure on machines at time t.

Energy costs p_2 per unit, and the second term is the cost of energy for all machines operating at time t.

We require that:

$$\int_{t-T(t)}^{t} Y(k)\,dk - Y^*(t) \geq 0 \tag{7}$$

The relation (7) states that the production from all machines must meet the output goal. Furthermore, research can increase the productivity of machines and reduce their energy requirements.

$$dA_1/dt = a_1\theta R^*(t), \quad dA_2/dt = a_2(1-\theta)R^*(t)$$
$$0 \leq \theta \leq 1 \tag{8}$$

where θ designates the portion of research funds devoted to each input factor.

$$\tau \leq 1 \quad \text{and} \quad \lambda + 1 \geq 0 \tag{9}$$

(9) implies that older machines are taken out before newer machines.

The first step in the analysis incorporates (7) into the objective function. This requires the use of the Lagrange multiplier L. The problem is also changed from minimization to maximization and restated below:

$$\mathop{Max}_{Y,\tau,\theta} \int_{t^*}^{\infty} e^{-r(t-t^*)} [-p_1(Y/A_1) - p_2\int_{t-T(t)}^{t} (Y/A_2)\,dk +$$
$$L\int_{t-T(t)}^{t} Ydk - LY^*]\,dt = -J \tag{10}$$

As stated in (10) the outer integral sums over time while the inner integrals sum over vintages operating in a given year.

127

An equivalent statement is possible with the outer integral being over vintages and the inner over years of operation. The equivalent form is given by

$$
-J = \begin{matrix} Max \\ Y, \lambda, \theta \end{matrix} \int_{t^*-T(t^*)}^{\infty} e^{-r(t-t^*)} \left[-\Omega(t) (Y/A_1) p_1 - (Y/A_2) \int_{t}^{t+M(t)} \Omega(x) e^{-r(x-t)} \right.
$$
$$
p_2(x) dx + Y \int_{t}^{t+M(t)} \Omega(x) e^{-r(x-t)} L(x) dx - \Omega(t) L(t) Y^* \right] dt \quad (11)
$$
$$
= \int_{t^*-T(t^*)}^{\infty} e^{-r(t-t^*)} F(Y, A_1, A_2, M) dt
$$

The function $\Omega(t)$ gets rid of costs previous to t^*. It is defined by

$$
\Omega(t) = \begin{cases} 1, t \geq t^* \\ 0, t < t^* \end{cases}
$$
$$(12)$$

To form the current value Hamiltonian (V. L. Smith, 1977) we use the adjoint variables η, μ_1, and μ_2:

$$
H = F(Y, A_1, A_2, M) + \eta \lambda + \mu_1 a_1 \theta R^* + \mu_2 a_2 (1-\theta) R^*
$$
$$(13)$$

128

4.10   THE ANALYSIS OF NECESSARY CONDITIONS

We derive the necessary conditions of the above vintage model and provide for each a suitable economic interpretation in the context of the proposed model.

For $t \geq t^*$ we have

$$\int_t^{t+M(t)} e^{-r(x-t)} L(x) \, dx - \frac{p_1}{A_1} - \frac{1}{A_2} \int_t^{t+M(t)} e^{-r(x-t)} p_2(x) \, dx \leq 0 \qquad (1)$$

Equality holds in (1) if Y(t) is greater than zero. L can be interpreted as the current marginal value of a new unit of production at time t. When Y(t) is greater than zero, the discounted value of a new unit of production over a machine's lifetime, must equal the capital cost per unit of production plus the discounted lifetime energy costs per unit of production.

$$\eta \leq 0 \qquad (2)$$

η is the value of retiring newer units before old. A closer examination of the problem shows that η always equals zero. The nature of technical progress, as described in 9(8), assures that A_1 and A_2 never decrease. New technologies are better in every way than old; therefore, a newer technology never will be retired before an older. (If new machines were improved in one characteristic but worse in another, this would not necessarily be the case).

$$\begin{aligned}
(\mu_1 a_1 - \mu_2 a_2) R^* &< 0, \quad \theta = 0 \\
(\mu_1 a_1 - \mu_2 a_2) R^* &= 0, \quad 0 \leq \theta \leq 1 \\
(\mu_1 a_1 - \mu_2 a_2) R^* &> 0, \quad \theta = 1
\end{aligned} \qquad (3)$$

μ_1 and μ_2 can be interpreted as the values of improvements in capital and energy use, respectively. (3) assures that research is devoted to the input which provides the greatest value, and that research is only done on both inputs when equally valuable.

$$\frac{d\eta}{dt} - r\eta + e^{-rM(t)} Y(t) \left[L(t+M(t)) \frac{p_2(t+M(t))}{A_2(t)} \right] \leq 0 \qquad (4)$$

Equality holds in (4) if M is greater than zero. M equal to zero practically does not occur, because purchasing and never using a machine cannot be optimal.

$$\frac{d\mu}{dt}\,r\mu_1 + \frac{p_1}{(A_1)^2}\,Y = 0 \tag{5}$$

$$\frac{d\mu_2}{dt}\,r\mu_2 + \frac{Y}{(A_2)^2}\int_t^{t+M(t)} e^{-r(x-t)}\,p_2(x)\,dx = 0 \tag{6}$$

(5) and (6) state that the values of research on capital and energy use decrease at the social discount rate and increase with the marginal value of an improvement in the technical coefficient. Since η equals zero, $\dfrac{d\eta}{dt}$ equals zero and (4) can be simplified and rewritten as

$$L(t+M(t)) = \frac{p_2(t+M(t))}{A_2(t)} \quad \text{or} \quad L(t) = \frac{p_2(t)}{A_2(t-T(t))} \tag{7}$$

L(t) is the current marginal value of a new unit of production. (7) states that a machine is retired when its operating cost per unit of production equals the current marginal value of a new unit of production. L(t) is an important quantity; but to investigate it further, additional assumptions are required. If Y(t) is greater than zero, (1) holds with equality and can be restated as:

$$\int_t^{t+M(t)} e^{-rx}L(x)\,dx - \frac{p_1}{A_1}e^{-rt} - \frac{1}{A_2}\int_t^{t+M(t)} e^{-r(x+t)}\,p_2(x)\,dx = 0 \tag{8}$$

By taking the time derivative of (8) and using (7) to simplify the result, we derive

$$L(t) = \frac{1}{A_1}\left[p_1(r+A_1^*) - \frac{dp_1}{dt}\right] + \frac{1}{A_2}\left[p_2 + A_2^*\int_t^{t+M(t)} e^{-r(x-t)}\,p_2(x)\,dx\right] \tag{9}$$

(9) simply states that the marginal value of a new unit of production must equal its marginal cost.

The first parenthesis on the right-hand side is divided by A_1 to put capital related costs on a per unit of output basis. The first term within this parenthesis shows the capital cost being amortized at the discount rate plus the rate of capital improvement. The next term is the capital savings lost by purchasing now rather than later. The second pair of large parentheses contain energy related costs and are placed on a per unit basis by dividing by A_2. Inside are the current energy cost plus the savings in lifetime energy costs which are lost by purchasing now rather than later. An interesting point is that L(t) does not depend on $Y^*(t)$, except that $Y^*(t)$ must be great enough since $Y(t)$ is greater than zero.

Necessary conditions (3) - (6) determine the research balance. Here we examine the case in which both technical coefficients improve. When this occurs the value of research in capital and energy must be equal over a period of time. If $\mu_1 a_1$ equals $\mu_2 a_2$ over time, then the change of each with time must also be equal:

$$a_1 \frac{d\mu_1}{dt} = a_2 \frac{d\mu_2}{dt} \tag{10}$$

Using (5) and (6) this condition is met only if a_1 times the marginal change in costs when A_1 is improved equals a_2 times the marginal change in costs when A_2 is improved.

$$a_1 \frac{p_1}{(A_1)^2} Y = a_2 \frac{Y}{(A_2)^2} \int_t^{t+M(t)} e^{-r(x-t)} p_2(x)\, dx \tag{11}$$

(11) is perhaps the most important derivation in this part, and has significant implications. It simply states that the value of research in each input must be equal if we commit research funds to each, a_1 and a_2 measure the effectiveness of research in improving the efficiency of capital and energy, respectively.

The marginal change in costs when the efficiency of an input is improved measures the value of changing A_1 and A_2.

(11) can be rearranged to give an expression in terms of the ratio of the technical coefficients. The ratio of the technical coefficients is determined by the ratio of the lifecycle costs times the ratio of the research efficiencies.

$$\left[\frac{A_2}{A_1} \right]^2 = \frac{a_2}{a_1} \int_t^{t+M(t)} e^{-r(x-t)} \frac{p_2(x)}{p_1}\, dx \tag{12}$$

4.11 NEUTRAL TECHNICAL PROGRESS

Improving the performance of the economy while not changing the ratio of the technical coefficients is the Leontief technology equivalent of neutral technical progress. The following proposition underlines the existence of such a ratio:

Proposition 1

Given the vintage model 9(1) to 9(9), if p_1 and p_2 are constant and $R^*(t)$ equals $R^* \exp(\gamma\, t)$ there is a unique ratio of technical coefficients A_1^* to A_2^* such that θ and M are constant along an optimal path. The ratio is expressed in

$$\left[\frac{A_2^*}{A_1^*}\right]^2 = \frac{a_2}{a_1} \frac{p_2}{p_1} \frac{(1-e^{-rM})}{r} \tag{1}$$

Proof

If A_1 and A_2 are invested in at a constant ratio, each will change at the rate γ. By substituting from 10(7) into 10(9):

$$e^{\gamma T} = \frac{A_1}{A_2} \frac{p_1}{p_2} (r+\gamma) + 1 + \gamma \frac{(1-e^{-rM})}{r} \tag{2}$$

This equation can be solved for the ratio of A_2 to A_1 and substituted into 10(12) to yield.

$$\left[\frac{p_2}{p_1} \frac{1}{(r+\gamma)} \left(e^{\gamma T}-1-\gamma\frac{(1-e^{-rM})}{r}\right)\right]^2 \frac{a_2}{a_1} \frac{p_2}{p_1} \frac{(1-e^{-rM})}{r} = 0 \tag{3}$$

Assuming M is constant, M equals T. At M equals zero, the left-hand side of (3) is zero and the first derivative of the left-hand side with respect to M is negative. At M equals infinity, the left-hand side of (3) is infinite. The second derivative of the left-hand side with respect to M is positive for all values of M greater than zero. From this we can conclude that there is a single, strictly positive value of M which satisfies (3).

Once the lifetime which satisfied (3), M*, is found, (3) can be solved for the ratio of A_1 to A_2.

To prove that the necessary conditions are also sufficient, we would need to first show that $F(Y, A_1, A_2, M)$ in 9(11) is concave in both control and state variables. We have examined the matrix which determines concavity of $F(Y, A_1, A_2, M)$, and found that

132

concavity depends on parameter values.

4.12 INDUCED PRICE CHANGES

The existence of a CO_2 problem adds on additional cost to the use of fossil fuels. In previous models this is captured through a shadow price, q. In this analysis we examine the response of the above equilibrium to an increase in the price of fossil fuel, which might be caused by the CO_2 induced shadow price. More precisely, we examine an increase in the ratio between the fossil fuel and capital prices. We find that an increase in the relative price of an input results in increased research in that input and a corresponding fall in the purchases of that input relative to the other input. The calculation is complicated because the optimal machine life changes when prices change. We first derive a relationship between the change in the equilibrium ratio of technical coefficients, A_2^*/A_1^*, to the change in the constant machine lifetime, M^*. Rearranging terms in 11(1), we derive

$$\frac{p_1}{A_1^*} = \frac{a_2}{a_1} \frac{A_1^*}{A_2^*} \frac{p_2}{A_2^*} \frac{(1-e^{-rM^*})}{r} \tag{1}$$

If we substitute from (1) and 10(7) into 10(9) and rearrange terms:

$$e^{\gamma M^*} = \frac{a_2}{a_2} \frac{A_1^*}{A_2^*} \frac{(1-e^{-rM^*})}{r} \; (r+\gamma) + \left[1 + \gamma \frac{(1-e^{-rM^*})}{r} \right. \tag{2}$$

We further rearrange (2) so that A_1^*/A_2^* and M^* each appear in separate terms.

$$\frac{a_2}{a_1} \frac{A_1^*}{A_2^*} \; (r+\gamma) \; + \; \gamma \; - \; r \frac{(e^{\gamma M^*}-1)}{(1-e^{-rM^*})} \; = \; 0 \tag{3}$$

By taking the total derivative of (3), we can find the direction of change in the optimal lifetime when the optimal ratio of technical coefficients changes. The total derivative is:

$$rG/[1-e^{-rM^*}]^2 dM^* - \frac{a_2}{a_1} \; (r+\gamma) \; d\left[\frac{A_1^*}{A_2^*} \right] = 0 \tag{4}$$

G is defined by

$$G = \gamma e^{\gamma M^*} + r e^{-rM^*} - (\gamma + r) \, e^{(\gamma - r) M^*} \tag{5}$$

At M^* equal to zero, G is zero and the derivative of G with respect to M^* is positive;

134

therefore, G is positive. From (4) we can then conclude that the optimal ratio of technical coefficients and the lifetime change in the same direction.

We complete the analysis of the impact of price changes by taking the total derivative of 4(1). The total derivative is

$$d\left[\frac{p_1}{p_2}\right] - \left[2\frac{a_2}{a_1}\frac{A_1^*}{A_2^*}\frac{(1-e^{-rM^*})}{r}\right] \quad \text{x} \quad d\left[\frac{A_1^*}{A_2^*}\right] - \left[\frac{a_2}{a_1}\frac{A_1^*}{A_2^*}\right]^2 e^{-rM^*}dM^* = 0 \qquad (6)$$

The multipliers of $d(A_1^*/A_2^*)$ and dM^*, the changes in the ratio of technical coefficients and the lifetime respectively, have the same sign. From (6) we then can conclude the following rule: when the ratio of the equipment price, p_1, to the fossil fuel price, p_2, increases, the level of research in capital relative to fossil fuel increases. Conversely, when the fossil fuel price goes up, relatively more research is being done on the efficiency of fossil fuel use. The research (and development) allows the economy to substitute equipment for energy in production.

4.13   CONSIDERATIONS OF MODEL WORKABILITY

Up to this point we have been concerned with model responses to specific patterns of shadow price increases. But it is difficult to examine even simple changes with this model. We look for an equilibrium when carbon dioxide causes the fossil fuel price to rise along an approximately exponential path. We find that the research response in this model to such a price rise is complex and cannot be described in simple terms.

It is natural to examine three possible equilibrium investment patterns: (i) balanced investment, (ii) all investment in A_1, and (iii) all investment in A_2.

If we examine the equation governing the balanced growth path, 10(12), we see that a balanced growth path with constant lifetimes and exponentially increasing fossil fuel costs is not possible. According to 10(9) and 10(7) lifetimes cannot vary in any simple way so that a balanced growth path is possible. Putting all our research in A_1 also cannot satisfy the necessary conditions. With all research in A_1, 10(9) and 10(7) give two different expressions for L. Neither can all research in A_2 produce a simple solution. From 10(9), when all investment is in A_2, the lifetime can only be constant if the rate of increase in research and the price is the same. If these rates are the same and the lifetime is constant, μ_1 is constant and μ_2 decreases at an exponential rate. The value of research in capital must eventually become higher than the value of research in fossil fuel. If all research is in A_2 and the lifetime is not constant, we cannot show that the necessary conditions are not satisfied, but any solution would be quite complex.

In order to accommodate some of these limitations, we will change the model in two important ways.

First, we assume that energy use is proportional to the number of machines in use and *not* the level of production. In the previous model, if no research was done on energy use but if machines were made twice as productive, each machine would consume twice as much energy. We now move to a situation, where machines may be made more productive without increasing the energy consumption per machine. In this modified model, the energy use of each vintage is:

$$F(k) \; - \; I(k)/A_2(k) \tag{1}$$

Second, we assume that over time the rate of growth in research funds, R, and the price of fossil fuels, p_2, is the same. This seems to be reasonable if the research budget and the harm due to CO_2 are both proportional to an exponentially growing economy.

We conclude that given the above assumptions, technical progress is neutral when the cost of energy rises exponentially. In this regard, the model appears less realistic than the previous model. However, this model's behaviour may give an indication of the performance of similar, more realistic, but more complex models.

4.14 SPECIFIC MODEL CHARACTERISTICS

The necessary conditions are quite similar to those of the previous model. The major difference is that the output per unit of energy input is the product of A_1 and A_2. The Lagrangean equations are replaced by

$$L(t+M(t)) = \frac{p_2(t+M(t))}{A_1(t)A_2(t)} \tag{1}$$

$$L(t) = \frac{1}{A_1(t)}\left[p_1(r+A_1^*) - \frac{dp_1}{dt}\right] + \frac{1}{A_1(t)A_2(t)} \tag{2}$$
$$\left[p_2 + (A_1^* + A_2^*)\int_t^{t+M} e^{-r(x-t)} p_2(x)\,dx\right]$$

(1) states that a machine is retired when the value of production from a new machine equals the operating cost per unit of production of a machine to be retired. (1) can be interpreted just as 10(7). L is again the present marginal value of a new unit of production.

The first pair of parentheses on the right-hand side of (2) contains the capital costs per unit of output, and the second pair contains the energy costs per unit of output.

Proposition 2

Given the energy requirement in 13(1), if p_1 is constant and p_2 and $R^*(t)$ grow at the rate γ, there is a unique ratio of technical coefficients, A_1^* to A_2^*, such that θ and M are constant along an optimal path.

The ratio is expressed as:

$$\frac{A_2^*}{A_1^*} = \left[\frac{a_2}{a_1} - \frac{A_2^*}{A_1^*}\right]\frac{p_2(0)(e^{(\gamma-r)M^*}-1)}{A_2(0)p_1} \tag{3}$$

Proof

The proof is very similar to Proposition 1. (3) is derived by equating the rates of change of the value of research in each technical coefficient. The equation which allows us to solve for the constant lifetime, M^*, is:

$$e^{2\gamma M^*} = \frac{A_2(0)}{p_2(0)}p_1(r+\gamma) + \frac{2\gamma}{r}(1-e^{-rM^*}) + 1 \tag{4}$$

These two models illustrate the difficulties under which the relationship of energy to

138

production is specified in order to determine the proper research response to increasing CO_2.

4.15 CONCLUSIONS

By referring to the step model of CO_2 impacts we observe that in such a model a critical CO_2 level imposes a long-run limit on fossil fuel use, and this limit is reflected in the shadow price of fossil fuel.

We assume that fossil fuels have no cost except for the shadow price.

In our reference model the level of output was constant and exogeneously specified. The level of fossil fuel use continually falls along an exponential path as efficiency in fuel use is gained. Therefore, in such a model, long-run limitations on fossil fuel use are irrelevant.

In the two models presented here such limitations are relevant. Let us analyse this property more closely.

In the first model, demand Y^*, and the research budget, R^*, grow at the same rate.

The solution is very similar to the initial model presented. We gain one unknown in the shadow price, but the added condition that emissions be less than or equal to the absorption rate times the critical CO_2 level allows us to solve for this unknown. A balanced research policy with a fixed ratio of knowledge about capital and fossil fuel use satisfies the necessary conditions. The shadow price in this model is constant.

In the second model rather than meeting a set production goal, we use the utility maximization framework. The constant relative risk aversion utility function is assumed. Output can only be raised by adding additional machines, there is no research on improved machine productivity. The level of research on the efficiency of fossil fuel use is an endogenous variable. Maximization is over the investment in new machines, the level of research on fossil fuel use, and the machine lifetimes. The economy, consumption, production, investment, research, all grow at a single internally determined rate. The shadow price rises at the rate of growth of utility which is the growth rate times one minus the consumption elasticity of utility.

The main conclusion of this part of the paper is that economic factors may have dramatic effects on the pattern of technical development. We have found that the neutral pattern of technical development, often assumed in energy environmental economic studies can be the outcome of several different sets of economic assumptions. A second major conclusion is that the type of impact is highly dependent on which model structure is considered most appropriate.

140

REFERENCES

Dixit, A. K., *The Theory of Equilibrium Growth*, Oxford University Press; Oxford, 1977.

Dixit, A. K., 'Growth Theory after Thirty Years', in P. Diamond (ed.), *Growth/Productivity/Unemployment*, MIT Press: Cambridge, Mass 1990, pp. 3-22.

Goodwin, H., *Chaotic Economic Dynamics*, Clarendon Press, Oxford 1990.

Ingham, A., Ulph, A. et al., "A Vintage Model of Scrapping and Investment", University of Southampton, Discussion Papers in Economics and Econometrics, No. 8724, Nov. 1987.

Jorgenson, D.W., "Capital as a Factor of Production", in Jorgenson, D.W. and R. Landan (eds.), *Technology and Capital Formation*, MIT Press: Cambridge, Mass., 1989, pp.1-35.

Manne, A. S., and R. G. Richels, *Buying Greenhouse Insurance: The Economic Costs of CO_2 Emission Limits*, MIT Press: Cambridge, Mass. 1992a.

Romer, P. M., "Endogeneous Technological Change", Journal of Political Economy 98, 1990, 71-102.

Schelling, T. C., "Economic Responses to Global Warming", in *International Burden Sharing and Co-ordination: Prospects for Co-operative Approaches to Global Warming*, Brookings: Washington D.C., 1991.

Scherago, J. D. et al, *GEMINI: An Energy-Environmental Model of the United States*", A Status Report on Model Development, Workshop on Economic/Energy/ Environmental Modelling for Climate Policy Analysis, University of Tokyo/MIT Centre for Energy Policy Research, Oct 22-23, 1990.

Smith, V. L., "Control Theory Applied to Natural and Environmental Resources", *Journal of Environmental Economics and Management* 4, 1977, 1-24.

Smith, V. Kerry, *Technical Change, Relative Prices and Environmental Resource Evaluation*, Resources for the Future, Johns Hopkins University Press: Baltimore, 1974.

Solow, R. M., *Capital Theory and the Rate of Return*, F. de Vries Lectures, North Holland: Amsterdam, 1964.

Solow, R. M., *Growth Theory: An Exposition*, Oxford University Press: Oxford, 1988.

Spash, C. L. and d'Arge, R. C., "The Greenhouse Effect and Intergenerational Transfers", *Energy Policy* 17(2), April 1989, 88-96.

Uzawa, H., "Optimal Growth in a Two-Sector Model of Capital Accumulation", Chapter 17 in H. Uzawa, Preferences, Production and Capital, Cambridge University Press: Cambridge, 1988.

CHAPTER 5

ENERGY - ECONOMY - ENVIRONMENTAL MODELS WITH SPECIAL REFERENCE TO CO_2 EMISSION CONTROL

5.1 INTRODUCTION

As a consequence of the modelling exercise in previous Chapters 2-4 the question arises as to how these models could be extended and improved for large- scale use so as to analyse more detailed, more specific policy options.

Before we suggest extensions, generalizations and ramifications of such models for more policy-directed use we review the state of the art of energy modelling today and possibly derive the lessons that can be learnt to date. As a natural starting point we examine the world energy and carbon dioxide model by Edmonds and Reilly (1983) and the model by Nordhaus and Yohe (1983).

5.2 REVIEW OF ENERGY - ECONOMY - ENVIRONMENTAL (EEE) MODELS

We will start with some early modelling approaches of W. D. Nordhaus (1979, 1980), in the context of carbon dioxide policies, involving an activity analysis model adapted to the CO_2 problem.

This is an instructive example of the application of simple optimization models to a quantitative, qualitative and integrated analysis of CO_2 strategies. The analysis contains the major ingredients of EEE models of this kind:

(i)   the dynamics of the CO_2 cycle, the sources of CO_2 and the diffusion of atmospheric CO_2 and the limits of CO_2 concentrations;

(ii)   the CO_2 energy model, e.g. a multi-sector activity analysis model which involves step-wise linear programming type optimization over a set of equidistant periods (ten periods each with twenty years);

(iii)   the development of control strategies based on shadow prices of CO_2 emissions and costs of abatement.

There are some obvious links between this approach and the models presented in Chapters 2 and 3: an explicit statement of an objective function and an equation for change in atmospheric CO_2. Nordhaus' consumption function includes two terms, a term dependent on fossil energy use alone minus a term dependent on CO_2 alone. Of particular interest is Nordhaus' formulation of a 'carbon tax' on fossil fuel emissions, as an outcome of deriving CO_2 control strategies. Otherwise, his study appears to be more narrowly focused, and since his model is not truly dynamic he does not address issues such as uncertainty, various types of technological progress, intergenerational equity and international co-operation, or changes in capital formation.

A new generation of global EEE (Peck and Teisberg, 1992; Nordhaus, 1993) models contains special features, e.g. the explicit consideration of the dynamics of economy and climate and a real interactive link-up between economic and climatic dynamics models, relating to a time path of global mean temperature. Nordhaus (1993, p.28) mentions that 'the earlier studies had a number of shortcomings, but one of the most significant from an analytic point of view was the inadequate treatment of the dynamics of the economy and the climate'.

A much more detailed and more extensive EEE modelling effort has been initiated by

Edmonds and Reilly (1983, 1985). However, there are still weak points. The model is not optimizing, feedback effects are not considered, capital investment is not traced and uncertainty is neglected.

The model matches the demand and supply for energy at discrete points in time in individual world regions. The model structure is relatively simple, but its disaggregation of data both increases realism and makes the results difficult to track. The model considers nine world regions, six major primary fuel categories, and three energy demand sectors. Energy demand is first projected based on population, economic activity, technological change, energy prices, and energy taxes and tariffs.

The model determines the supply of primary energy in three different ways. Resource constrained fuels, such as conventional oil and gas, are produced according to resource depletion curves and their supply is unresponsive to price. Hydro power and biomass are considered resource constrained renewable sources which reach and maintain a specified production level. Their production is also insensitive to price. Finally, sources such as coal, nuclear, solar and unconventional oil and gas are classified as unconstrained. The production levels of these sources are not limited but depend on price. The primary fuels are converted to secondary fuels and then to energy services. The prices of energy services depend on the price of primary energy, the mix of energy sources providing the service, and transportation and conversion costs. The initial fuel mix is exogenous, but the mix is then modified in light of energy prices. The final energy use is determined by an iterative process which changes prices, balances energy trade among regions and balances energy supply and demand for each region. As energy prices vary, demand and supply adjust in response to exogenously specified demand elasticities and supply functions. Based on the mix and level of fossil fuel use, the carbon dioxide emissions are calculated. The retention of carbon dioxide in the atmosphere and the consequent climate changes are determined by a sophisticated model which considers other greenhouse gases, ocean absorption of CO_2 and other factors.

The Edmonds and Reilly (ER) model is part of an integrated U.S. EPA report (1983). The report's major conclusions are that little can be done to delay a greenhouse warming and that major uncertainties include the effects of other greenhouse gases and the temperature sensitivity of the atmosphere. The report emphasizes the importance of coal use in determining the CO_2 level and of the trace gases in determining the eventual temperature increase. In the ER model, logistic curve shaped paths are exogenously specified for several

145

new energy technologies. A related MIT study, Rose et al (1983), modifies these paths and adds additional technologies. It finds that the adoption of realistic but CO_2 benign technologies, while not eliminating a significant CO_2 warming, could increase the CO_2 doubling time to several centuries.

Another major and comprehensive report is that of the National Research Council (1983). It uses energy-economy, climate and agricultural models to predict future impacts of carbon dioxide and trace gas accumulation. The major conclusions are that no radical action should be taken, that increases in carbon dioxide are likely, which will cause measurable rises in the world's average temperature, and that more research is necessary.

Referring to the energy-economy model by Nordhaus and Yohe (NY), contained therein, one can note that the technology development and elasticity of substitution parameters critically affect the model's results. More specifically, the NY model maximizes consumption in individual time periods. A highly aggregate function determines production - the model considers no world regions and only two fuel types, carbon and noncarbon. The model is not optimizing over time, does not include feedback effects, and does not consider capital stocks. The most significant characteristic of this model is the estimation of probability distributions for important future variables.

Furthermore, overall technical progress is assumed at a constant rate. The production function relates the energy and labour sectors in a Cobb-Douglas fashion; however, a modified form of this function is used. In the usual Cobb-Douglas formulation, the production elasticity of an input is constant and equal to the share of GNP assigned to the input. Further, in the usual formulation the elasticity of substitution between inputs is always -1. These rigidities are a major drawback to the use of the Cobb-Douglas function.

In the NY model the payments to inputs are adjusted in each time period. The adjustment is consistent with a changing, exogenously set elasticity of substitution between labour and energy. The reason is to overcome the limitation placed on substitution elasticities by the Cobb-Douglas function. Within the energy sector there is a constant elasticity of substitution between the fossil and non-fossil inputs.

The price of noncarbon based fuel equals the sum of distribution costs and production costs. The distribution costs are constant, but the production costs change exponentially over time. The rate of change is the sum of a term representing the technical change in the energy industry and a term representing a bias towards noncarbon energy.

146

The equation for carbon fuel prices is similar but somewhat more complex. The first term combines both production costs and costs due to fuel depletion. Further, the second term is multiplied by the exponential change in energy industry technology, but there is no bias term. Finally, a tax on carbon fuels may be added. The emissions of carbon into the atmosphere per unit of carbon fuel is assumed to grow over time, because the mix of fuels includes more high carbon content fuels.

They assume that parameters describing the current economy can be determined accurately, although they wish to account for an inherent uncertainty in future values of parameters.

The authors estimate the probability distributions on important parameters using the distribution of published predictions.

A culmination of efforts in the category of global EEE models has been Nordhaus (1993), (1994). His approach centres around the construction, integration, model assessment and policy analysis of his economic control model DICE (Dynamic Integrated Model of Climate and Economy). DICE is a dynamic, intertemporal, optimal, interactive, welfare-economic control model based on structural equation constraints such as population growth, production constraints, capital stock accumulation, emission constraints (where GHGs are normalized by their carbon dioxide equivalent in terms of their "global warming potential" (GWP)). The model contains a critical economy climate interface that links GHG emissions to their accumulation and transport in the atmosphere, the radiative forcing of the GHGs and their links to climate change. To assess the economic impacts such as damages, DICE contains feedbacks from climate change to economics, by specifying the loss of global output due to climate change. The climate part of DICE relates to specifications of General Circulation Models (GCMs), condensed as a "minimodel" of climate change to have it fit with the economic interface. Given the structure of DICE Nordhaus puts his model to test. First he estimates damage profiles of GHG induced damages, for particular sectors as well as enticing the entire GDP loss. Furthermore, he looks at the welfare economic implications (net benefits) of seven major policy strategies to control global climatic changes: (1) no controls, (2) optimal policy, (3) ten year delay of optimal policy, (4) stabilizing emissions at 1990 rates, (5) 20 percent emission reduction from 1990 levels, (6) geoengineering, (7) climate stabilization with upper limit of total mean temperature increase by 1.5 C from 1990. The net benefits vary significantly in size from each other, where it is remarkable that, in

147

general, more interventionistic strategies (stabilization), as strongly advocated by environmentalists, fare much worse than less interventionistic ones (except for geoengineering which, of course, is hardly to the environmentalist's delight).

The extent of uncertainty in the model parameters gives rise to estimating the impact range on strategic outcomes as well as it applies to regulatory decision making on how to optimally impose regulatory controls to minimize over or undershooting of environmental regulation and policy measures (the value of information of waiting vs acting). As further explored in Chapter 9, choosing the level of GHG emission limiting regulations that will maximize social welfare by optimally balancing the costs of emission control against the benefit of decreased environmental damage is inherently difficult because of pervasive uncertainty about the likely size of the critical GHG budget, its relationship to the quantity of GHG emitted, the effects of GHG in the atmosphere, and the appropriate valuation of these consequences.

5.3 TREATMENT OF NEW ENERGY TECHNOLOGIES

Because the levels of fossil and non-fossil fuel development will have enormous impacts on the severity of the CO_2 problem, we find that technical change and technology substitution are specified exogenously or modelled in a simple manner.

In some technology development models, for example Marchetti (1975), long development cycles over several decades are given for the introduction of significant levels of non-fossil fuels. In these models forecasts on the use of new fuels are based on their growth immediately following introduction and the assumption of a logistic growth path in the future. The logistic assumption is very popular with technologists because of its reasonable looking symmetric shape and good fit with the historical growth of new technologies.

Peterka (1977) and Spinrad (1980) consider the logistic process as very natural for the proliferation of energy technologies.

This approach, however, has two specific problems which we seek to avoid in our description of technology: the observations do not have a good theoretical basis and the technologies appear exogenously rather than through a process of research (Chapter 2).

In studies of CO_2 by Perry et al (1982) the authors first assume an energy demand level and a fossil fuel use pattern. Fossil fuel use follows a logistic curve between the present and an assumed ultimate level of use. The rate of non-fossil energy growth needed to fill the gap between fossil energy use and an assumed total energy demand is then examined. Their work emphasizes the importance of analysing the investment needed in non-fossil energy to fill this gap but does not present a model of the substitution process. In our models, the substitution process is a direct result of our maximization of welfare.

In these studies we see an agreement about the importance of technological progress but a lack of detailed modelling. None of them show the model-theoretic connection between economic factors and changing technologies developed in our models.

As a prototype model for treating technical progress and capital investment in energy-economy interactions we can consider the ETA-MACRO model by A. S. Manne (1977). This model can be described as a multi-sector 'look ahead' model simulating a market economy through a dynamic, non-linear optimization process. It examines consumption and investment policies and their impact on national welfare. National welfare is measured by discounting utility from the present to a distant future. ETA-MACRO consists of two models

- a macro-model of the whole economy and a more detailed model of the energy sector.

In this model future changes in energy technologies and input-output coefficients may be exogenously specified. It assumes that capital in place requires fixed inputs of energy, labour etc, but that new technology uses the most efficient combination of inputs. This is the 'putty-clay' assumption. The model assumes a single capital good can be used to help produce any type of energy. ETA-MACRO uses a single energy aggregate distributed to the rest of the economy.

The model seems to be more sophisticated in its treatment of capital and the determination of the desirable level of energy use than the models presently used in CO_2 analysis. However, in its present form it is unsuitable for examining international problems such as CO_2. It also lacks a treatment of endogenous technical progress.

Such a model has desirable features for EEE interactions. It is optimizing and considers costs and benefits of capital investment.

More recently, Manne and Richels (1990), Manne (1992b) have proposed substantial modifications of his ETA-MACRO model for CO_2-energy-economy interactions integrated into his Global 2100 model. Global 2100 is a multi-regional model in which each region, pursuing its own interests, is a contributor to global carbon emissions. Since each region is likely to pursue its own individual interests rather than the global welfare, such a model could be solved within a computable general equilibrium (CGE) framework. In the same vein, there have been multi-sector, multi-regional models initiated through the OECD (Burniaux et al., 1992), the EU (EC, 1995), commonly using a CGE methodology (Gottinger, 1998).

We examine the effects of technical change on the economy and CO_2 accumulation in both single and multiple state models. The simplest type of technical change is a finite or limited improvement in a technology.

The key difference in the single and multiple state variable models is the ability to control technical change. In the single state models technical change is exogenous; in the multiple state models, the rate of technical change is determined by decisions on investment.

Another interesting extension of ETA-MACRO, albeit in its impact assessment limited to the U.S., is the recently established GEMINI model (Scheraga et al, 1990).

5.4   ECONOMIC MODELS OF POLLUTION AND CONTROL

A different class of models originating in pollution and environmental economics deals with the intertemporal optimization of welfare functions under resource constraints relating shadow prices to optimal taxes. Unfortunately, to our knowledge, there has been no attempt so far to link these models to EEE models and thus provide more sophisticated explanations for policy responses to global environmental issues. The general static model of pollution considers both consumers and producers. The model is solved for the conditions which govern Pareto optimality. Baumol and Oates (1975) make clear the distinction between pollutants that impinge on utility rather than production. Many types of pollution affect utility directly such as air pollution; however, the major CO_2 impact will most likely affect production.

They also show that tax rates which achieve a desired reduction in pollutants will satisfy the necessary conditions for Pareto optimality. By reviewing optimal control models of pollution problems we observe that none of these models are appropriate for the analysis of CO_2.

Still there are some key elements that have to be integrated into general EEE models for global environmental problems.

In other respects, specific distinctions remain. For example, D'Arge and Kogiku (1973) introduced optimal control models with two important features: they assumed that pollution affected utility directly and their models consider a finite horizon. They found several non-intuitive results in their models, many caused by finite horizon assumptions. In the context of the issues pursued we are convinced that infinite horizon models make more sense. Three specific optimal control models, for different reasons, are of some interest for our modelling exercise.

Forster (1980) develops three models of pollution from a source which is in limited supply. In all three models, the pollution affects utility directly and acts as a flow. He is mainly concerned with the existence of equilibria. Asako (1980) and Becker (1982) are both concerned with issues of intergenerational equity and the contrast between maximin solutions and utility maximization. In models with technical progress, maximin solutions lead to higher current emissions than utility maximization (in a comprehensive sensitivity analysis for large-scale models other decision criteria could be used).

As a rough approximation this suggests that utility maximization with discounting is

151

not a particularly short-sighted approach to planning.

Another string of models and research results of resource economics (on the depletion of non-renewable resources) could be applied with simple modifications to the CO_2 problem.

Under two crucial assumptions the problem of fossil fuel use in the face of increasing carbon dioxide is parallel to the problem of consumption of a limited resource. The first assumption is that the carbon dioxide absorption rate is sufficiently small to be ignored. The second is that CO_2 impacts follow a 'step' pattern, that is, CO_2 (as a pollution stock) has no impact on productivity until a critical level, M_c, is reached; then if the CO_2 level exceeds M_c, production drops sharply (or more extremely, falls to zero).

A model with endogenous neutral technical progress, as in chapter 3, is proposed to provide a better explanation of technical changes used to date in EEE models. Such a model originates from a similar attempt by Chiarella (1980). He proves the existence of a steady state growth path and a simple rule governing the rate of investment in research. Research investment along the optimal path should be carried out until the growth rate in the marginal accumulation of technology equals the difference between the marginal product due to an extra unit of research investment and the marginal product of capital.

Another issue is uncertainty. Here again there is a link with models of resource use for a limited, non-renewable resource when the reserve of the resource is unknown.

The key finding of models by Loury (1978) and Gilbert (1979) is that plans for resource use based on the expected level of a resource will be overly optimistic. Gilbert's model is conducive to our models of fossil fuel use when the critical CO_2 level is uncertain. Under the above assumptions this problem is equivalent to determining the rate of fossil fuel use when the critical concentration of atmospheric carbon dioxide is unknown. Their results show that the optimal use of fossil fuel is lower when uncertainty is properly considered than when the expected values are assumed to be certainty equivalents.

A significant additional element is the possibility of undertaking exploration to find new reserves. The parallel in the CO_2 problem is R & D to increase the probability of finding a technology for the removal of CO_2 from the atmosphere.

Deshmukh and Pliska (1980, 1983) find that in the periods between discoveries or research breakthroughs, fossil fuel use and consumption fall, but if R & D is very successful, long-run fuel use may rise.

There are three major issues that have been addressed in our models but they have not

152

been dealt with in a systematic way in all the major EEE models proposed and implemented. These are international or multi-regional issues, technology issues and issues of uncertainty and risk. In a paradigmatic way we go through each of those issues and offer some solutions in the context of our models. We are still far from offering an integrated view.

5.5 INTERNATIONAL ISSUES

Much of the work on international relations, trade and pollution deals with legal issues and the form of international control agreements. Mathematical modelling has concentrated particularly on balance of payments and terms of trade when a country controls pollution within its own boundaries. In modern textbooks on trade theory and international economics, pollution problems have been neglected (Dixit and Norman, 1980).

Transnational pollution involving adjacent nations where one is the polluter and the other receives the pollution has received little attention, and global pollution where all or many nations contribute has received even less (Baumol and Oates, 1975, Chapter 1, 16) but some recent work (Ulph, 1990; Barrett, 1992) points to the right direction.

The difficulties just mentioned suggest that a single country or a group of countries may wish to unilaterally alter the international use of fossil fuels through export and import controls such as taxes and subsidies. We approach this problem as part of the modelling effort in Chapter 2. The model assumes there is a shadow cost, q, associated with the use of fossil fuels. The results are equally applicable if this is a static cost or a changing cost in a dynamic model. In the model the government wishes to control two variables, F, domestic fuel use, and F_e, exports of fossil fuels. Domestic production of fossil fuels, F_d, equals F + F_e. First assume U(C) equals C and that atmospheric CO_2 does not affect the economy (q = 0):

$$C = f(F) - c(F_d) + p_i F_e \tag{1}$$

f is the goods production function, c is the fossil energy cost function, and p_i is the international price of fossil fuels. In particular cases p_i may be a function of F_e. The two necessary conditions for maximizing consumption are:

$$f' = c' \tag{2}$$

$$c' = p_i + \frac{\partial p_i}{dF_e} F_e \tag{3}$$

In a competitive international market the partial derivative of p_i with respect to fossil fuel exports is zero, and the national government has no control over the international price, p_i, or the international use of fossil fuels. Production occurs until the marginal cost equals the international price.

Now consider a situation in which CO_2 accumulation has an adverse impact, but the impacts are not observed outside the country of interest.

Let F_i be the international fossil fuel consumption. Consumption can now be expressed as below.

$$C = f(F) - c(F_d) + p_i F_e - q(F + F_i)$$ (4)

q is the domestic shadow price of world, domestic plus international, fossil fuel use.

The necessary conditions can be stated as:

$$f' = c' + q$$ (5)

$$c' = p_i \frac{\partial p_i}{\partial F_e} \left[F_e - q \frac{\partial F_i}{\partial p_i} \right]$$ (6)

[Maximization of (4) gives the same necessary conditions as the optimal control problem with an export sector, except for the import difference that we have no information on the magnitude and rate of change in q. Because we only use the fact that q is positive, we can use the much simpler problem statement in (4).]

First, consider the competitive international case. Comparing (2) with (5) it is seen that f' is unequal in the two situations; therefore, domestic use of fossil fuel changes. A tax of q, which is greater than zero and changes over time in dynamic problems, can be placed on domestic fossil fuel use to reduce domestic use to the optimum. Since non-domestic (international) use does not change this will cause a decrease in world emissions. In the competitive case the partial derivative of the international price with respect to fossil fuel exports is zero. Equations (3) and (6) are identical in this case, therefore, domestic production and international use of fossil fuels are the same with or without CO_2 problems.

155

Because net exports equal domestic production minus domestic consumption, concern over CO_2 in a competitive environment has the odd effect of increasing exports of fossil based fuels.

A more realistic model would consider that even in a highly competitive market any increased exports would lower the world price, reducing world production and increasing consumption outside the concerned region. This more complex model, however, would not change the basic conclusion that, in a highly competitive market, exports from the concerned nation should increase.

Now consider the even more realistic case of a non-competitive world, a world in which the country or countries of concern are large enough producers of fossil fuels to affect the international price and use of fuels. In this case, changes in exports do alter the international price and consequently international consumption.

From (3) it can be concluded that even in the absence of CO_2 problems it will be optimal to tax exports at a rate equal to the partial derivative of the international price with respect to exports times the level of exports. This drives up the price and allows the domestic economy to take advantage of its quasi-monopoly position. When concern exists about CO_2 in a non-competitive market (6) shows that a still higher tax may be placed on exports. This tax further reduces domestic production and international fossil fuel use.

Equation (5) has the same form in both the competitive and monopoly cases: therefore, a domestic use tax may be used to achieve optimum domestic use. When markets are non-competitive we cannot determine if the reduction in domestic fossil fuel use is greater or less than the reduction in fossil fuel production; consequently, it cannot be determined if exports increase or decrease. Note that in a dynamic model, q will be different in competitive and non-competitive markets because of differences in the levels of emissions over time.

Another interesting case occurs when the country or countries concerned about CO_2 are major exporters of a fossil fuel substitute. If the market is very nearly competitive and the fossil fuel substitute is priced very close to fossil fuels, a small export subsidy may cause a very large switch to the substitute. As in the case of a fossil fuel export, if a monopoly position is realized there exists an incentive to tax exports of the fossil substitute and raise its international price. Because use of the substitute reduces CO_2 output, there may be opposing reasons to subsidize or tax international sales of the substitute. Only in specific

156

cases can it be determined whether the net result will be a subsidy or tax on exports. More sophisticated results have recently been obtained by M. Hoel (1990).

5.6 TECHNOLOGY ISSUES

The long-run level of atmospheric CO_2 depends on both the degree of technical change and its form. Technical change varies in its impact on the productivity of inputs and the reabsorption of CO_2.

In a simple case of a Cobb-Douglas technology we show the leverage of technical progress on the long-run use of fossil fuels.

In accordance with previously used notation the model assumes that a critical level of CO_2 accumulation, M_c, exists, at which production drops to zero. M_c is used to define two variables, F_s and M_s: $M_s = M_c - M$ and $F_s = F - \alpha M_c$.

M_s represents the distance from the critical level of CO_2 concentration and F_s is the distance from the level of emissions which maintains the critical CO_2 level. In the model consumption increases with increases in either of these variables, the consumption/production function is:

$$C = F_s^\epsilon M_s^{1-\epsilon} e^{\gamma t} \tag{1}$$

where $\gamma = (dS/dt)/S$ is the exponential rate of increase in 'knowledge' or technology S.

The necessary conditions are then stated as

$$\epsilon C^{-\beta} [M_s/F_s]^{1-\epsilon} e^{\gamma t} = q \tag{2}$$

where β is the consumption elasticity of utility.

$$\frac{dq}{dt} = (r+\alpha) q - (1-\epsilon) C^{-\beta} [F_s/M_s]^\epsilon e^{\gamma t} \tag{3}$$

The assumed solutions for F and M, i.e.

$$M_s = M_s(0) e^{gt}, F_s = F_s(0) e^{gt} = -(g+\alpha) M_s(0) e^{gt}$$

do not change from the static reference model, although the equation for consumption must be restated as

$$C = C(0) \, e^{(\gamma+g)\,t} = (-(g+\alpha))^{\epsilon} M_s(0) \, e^{(\gamma+g)\,t} \tag{4}$$

Examining equation (2) it is seen that

$$q = q(0) \, e^{(\gamma - \beta(\gamma+g))\,t} \tag{5}$$

By substituting for q and dq/dt in (3) and simplifying, we derive

$$\gamma - \beta \, (\gamma+g) = r + \alpha + \frac{(1-\epsilon)}{\epsilon} \, (g+\alpha) \tag{6}$$

Solving for g we get the following expression

$$-g = \frac{r\epsilon + \alpha - \gamma\epsilon \, (1-\beta)}{1 - \epsilon \, (1-\beta)} \tag{7}$$

Consumption grows at the rate $g + \gamma$. Using (7) we derive the following:

$$g + \gamma = \frac{\gamma - \alpha - r\epsilon}{1 - \epsilon \, (1-\beta)} \tag{8}$$

As noted in Chapter 2, for an optimum to exist the objective function must be finite. This requires that the growth in utility be less than the discount rate

$$(g+\gamma) \, (1-\beta) - r < 0 \tag{9}$$

Making the appropriate substitutions, (9) reduces to

$$\gamma - \alpha \frac{r}{1-\beta} < 0 \tag{10}$$

A high α results in a higher minimum level of fuel consumption, higher initial emissions, and a more rapid decrease in the use of fuels. A more rapid decrease in fuel use, in turn, means that the objective will be finite at higher levels of technical progress.

The condition in (10) does not reveal whether consumption is increasing or decreasing. (10) does assure that the numerator on the right-hand side of (7) is positive, therefore, if the optimum is finite, the use of fossil fuels is decreasing over time.

The model parameters affect the rate of change in energy use in the same fashion in the models with and without technical progress. Increases in the parameter, γ, reduce both initial energy use and the rate of decline in energy use. Equation (8) determines whether consumption increases or decreases over time. Technical progress tends to make consumption grow. Reabsorption, social discounting and high fossil fuel productivity all encourage high initial energy use and make a consumption decline more likely.

More specific conclusions from a Cobb-Douglas type technology can be drawn if we adopt a taxonomy of technologies to the control of CO_2 emissions, as discussed in Chapter 2, such as, for example, sequestering of carbon dioxide, development of non-fossil, substitutable energy, fossil fuel enhancing technical progress, amelioration of CO_2 impacts or emissions scrubbing. Other, more general technical progress functions could open more policy options [Ayres and Miller (1980)]. Such investigation, of course, would involve specific numerical models.

5.7 ISSUES OF UNCERTAINTY

Existing EEE models suffer from poor data, indeterminate structure, and a frequent lack of attention to the consequences of uncertainty. The factors linking energy activities to their environmental effects are known only imprecisely. EEE models rely on behavioral assumptions that are widely questioned, and on parameters that can vary substantially from one model or data source to the next.

Most EEE models, including those well-established and highly used, conceal this uncertainty behind a blanket of output detail: a profusion of fuel prices and quantities, sectoral disaggregation, regional detail, growth rates and target figures, which often steer the analysis toward a desired conclusion. Unfortunately, this complexity rarely contributes to a resolution of uncertainty, and may serve only to increase the error and expense.

Existing EEE models are quiet on the question of uncertainty.

Concerning models of possible greenhouse effects and CO_2 emissions, uncertainty analysis assumes many facets. It involves:

(i)   changes in climate to be expected

(ii)   impact of climate change

(iii)   costs of adapting to climate change

We distinguish between uncertainty about occurrences of events and impacts.

Policy uncertainty is also of great concern. For the CO_2 problem some argue that it is premature to think about doing other than intensive research, others claim that the risks of waiting are simply too great. What is the value of reducing scientific uncertainty? Scenario analysis only provides an indirect treatment of uncertainty, all uncertainties are resolved prior to decision-making.

But uncertainty, information and decision-making are intimately connected and a comprehensive approach based on Bayesian decision analysis shows promise (Manne and Richels, 1990).

As part of our optimal control model, Chapter 3, we consider structural uncertainty as affecting the model's parameters. Some structural uncertainty is an inherent property of complex EEE models. It relates to the stochastic nature of optimal control models (Holly and Hallet, 1989), adapted to our specific needs.

161

In such models the treatment of uncertainty of critical parameters by the use of expected values is common, but we show that anything less than a full treatment of uncertainty can lead to a biased calculation of the optimal present fossil fuel use.

To demonstrate how uncertainty naturally enters our model, as presented in Chapter 2, Section 12, we assume that the present level of CO_2 is $M_o = M(0)$ and that n distinct (possible) carbon dioxide levels, M_i exist. The prior probability that M_c equals M_i is π_i. J_i is the maximum expected value of future fossil fuel consumption when the current level of carbon dioxide in the atmosphere is M_i. p_{ij} is the updated probability that M_c equals M_j given that M_i has been reached and is not the critical level. E denotes an expectation, r is the discount rate. J_i is defined by

$$J_i = Max \ E \left[\int_0^\infty e^{-rt} U(C) \, dt \right] \tag{1}$$

such that

$$C = \begin{cases} f(F) , M \le M_c \\ 0 , M > M_c \end{cases} \tag{2}$$

$$M(0) = M_i \tag{3}$$

$$p_{ij} = P(M_j = M_c \ given \ that \ M_i \ne M_c)$$
$$= \begin{cases} \pi_k / \sum_{j=i+1}^{n} \pi_j , k = i+1, \ldots, \\ 0, otherwise \end{cases} \tag{4}$$

Let T_i be the time to move between CO_2 levels M_i and M_{i+1}.

Define the emission rate

$$\int_0^{T_i} e^{-\alpha(T_i-t)} F(t) \, dt = M_{i+1} - e^{-\alpha T_i} M_i \qquad (5)$$

with α being the CO_2 reabsorption coefficient.

Then J(M_i, M_{i+1}, T_i), the value collected moving from a CO_2 concentration of M_i to M_{i+1} in time T_i is defined as

$$J(M_i, M_{i+1}, T_i) = \underset{F}{Max} \int_0^{T_i} e^{-rt} U(C) \, dt \qquad (6)$$

such that

$$C = f(F) \qquad (7)$$

$$\frac{dM}{dt} = F - \alpha M \qquad (8)$$

$$M(0) = M_i \text{ and } M(T_i) = M_{i+1} \qquad (9)$$

Gilbert's analysis (1979) in a similar context is pertinent to our model. In order to facilitate optimization of J_i we could establish the algorithm

$$J_i = \underset{T_i}{Max} \left\{ J(M_i, M_{i+1}, T_i) + e^{-rT_i} \left[p_{i,i+1} \frac{U(C(\alpha M_{i+1}))}{r} + (1 - p_{i,i+1}) J_{i+1} \right] \right\} \qquad (10)$$

The solution algorithm (10) requires maximization over T_i of three terms. The first is the value gained while raising the CO_2 level from M_i to M_{i+1}. The second is the discounted value of consuming fuel at a rate that maintains the CO_2 level at M_{i+1} weighted by the probability that M_{i+1} is the critical CO_2 level. The third is the discounted, expected value of raising the CO_2 concentration above M_{i+1} weighted by the probability that M_{i+1} is not the critical level.

A further result is that a certainty equivalent critical level of carbon dioxide, M_{ce}, exists. When used in calculations of optimal emissions, M_{ce} is a certainty equivalent level of CO_2 which produces the same initial emissions as the algorithm in (10).

But within this framework of treating structural uncertainty we can show that M_{ce} is less than $E(M_c)$. It follows that the calculated optimal current fossil fuel use, when probabilities are fully treated, in terms of certainty equivalence formulation, is lower than the optimal current fossil fuel use, when expected values are treated as certain values.

Looking at particular examples, we compare two different treatments of uncertainty. In the 'expected case', the expected value of uncertain parameters is used in all calculations, and other aspects of uncertainty are ignored. In the 'base case', uncertainty is fully treated, that is, the certainty equivalence formulation is applied. If the step model of CO_2 impacts is used, Section 2.5, then changes in present fossil fuel use under the two treatments of uncertainty and a wide variety of parameter settings can be calculated. In the expected case suppose M_c equals M_1 with probability π and M_2 with probability $(1 - \pi)$. If the critical CO_2 level is assumed equal to the expected critical level

$$M_c = \pi M_1 + (1 - \pi) M_2$$

The problem is now 'certain', the present emissions level can be found, as in the step model. In the second approach to uncertainty, we make use of the probability estimates and the algorithm developed in (10). In this case, the atmospheric CO_2 certainly can increase from M_0 to M_1. $J(M_0, M_1, T)$, the value gained in the transition is certain, and calculated according to (10).

164

5.8 PHILOSOPHY OF EEE MODELLING

Because the feedback effects of CO_2 are extremely uncertain, many modellers are reluctant to incorporate these effects in their models. In some scenarios, feedback effects might indeed be unimportant. For example, in models with finite horizons, if CO_2 effects are insignificant until after the horizon of the model no modelling of feedback effects is needed.

In models which optimize over an infinite horizon, future effects may change current policies, and feedback effects are always of importance. However, in these same models feedback effects may make solution much more difficult.

In predictive models with long time horizons, feedback effects will also be important. Further, in predictive models that estimate production and energy use at individual points of time, such as those surveyed previously, feedback effects can be easily included.

Experience with the inclusion of feedback effects in optimizing models, as presented in Chapters 2 and 4, shows that they usually lower the optimal initial use of fossil fuels. The long-term changes in fossil fuel use due to feedback effects are more uncertain and dependent on the model. In general one could say that in most models the feedback slows the economy and thus reduces the demand for fossil fuels in the future. If optimization were included in the models discussed, this effect would be likely.

Given the uncertainty in the severity and timing of feedback effects, the sensitivity of individual models to variations in feedback effects is of much interest. In proposing a step model of CO_2 emissions we also examined the sensitivity of current fossil fuel use to an ultimate limit on atmospheric carbon dioxide. We found that current optimal fossil fuel use was significantly affected by different critical levels of CO_2. A study of the impacts of a critical CO_2 level in a more disaggregated optimizing model would be useful. Including optimization in models expands their applicability but may cause analytic problems and controversy. As with many social problems, an acceptable, objective function for carbon dioxide control problems is difficult to define. Any definition will seem both inadequate and overly precise and certainly will be controversial. This may be the reason why the models reviewed did not examine optimal policies. On the other hand, statement of an objective function does not hide or confuse other results and can add many new insights. If feedback effects and an objective function are included in a model, a crude optimization can be performed simply by running the model under a variety of policies.

Including optimization raised several new issues in our models. For example, pollution impoverishes but technical progress enriches the future. The optimizing models show how the curvature of the utility function, determined by the consumption elasticity of utility in our models, tends to smooth or even out wealth over time. Without an objective function being stated, the importance of this redistribution effect in determining fossil fuel use policy cannot be examined. In predictive models, a subjective evaluation must be made of the significance and value of a policy. In an optimizing model, the costs and benefits of policies are automatically compared in an explicit manner.

A final benefit of an optimizing model is the identification of multifaceted responses which may be ignored when policy changes are specified exogenously. Integrated optimizing models respond to problems by adjusting numerous policies endogenously. For example, in multiple state models fossil fuel use, research, and capital all respond to changes in the effects of CO_2.

One specific important factor is capital. The inclusion of capital could have a variety of effects. The explicit addition of capital would add greater flexibility to the models and thus could improve predictions of the future. Conversely the inclusion of capital may make certain high consumption growth paths look less attractive. If a high growth path requires a high proportion of GNP to be invested, GNP, the simplest measure of welfare, is no longer a good measure of the quality of the growth path. The models suggest that rapid growth is accompanied by very high investment rates.

In vintage models, such as the one proposed in Chapter 4, capital reduces the rate of change in the economy by requiring machine replacement to accompany changes in inputs. In the ER model this issue is explicitly discussed and their conclusion is that the time horizons are such that this effect is unimportant.

166

5.9 POLICY ANALYSIS OF EEE MODELS

Two structural features of the Edmonds and Reilly model deserve comment. Because the model is not forward looking, the addition of feedback effects and the definition of a welfare function should not be difficult and would provide the opportunity for some simple optimization of policies. Second, the treatment of technical change is unusual. Technical change is neutral across fuels and the rate of technical change declines over time. But neutral technical progress depends on special economic circumstances that cannot be generally expected. The declining rate of technical change is not common in other models we examined.

There are two substantial results of their study that stand out in particular. The first of these is the importance of the greenhouse gases. The authors state that much of the uncertainty regarding future temperature increases is due to these other greenhouse gases, and it is clear that, in part, CO_2 control policies are ineffective because they do not change emissions of other greenhouse gases. The models presented in Chapters 2 and 3 suggest that the worse the impacts of other greenhouse gases the lower the optimal present emissions of CO_2.

Other conclusions from the ER model are that it is very difficult to significantly reduce the temperature rise expected in the second half of the next century. This result appears to have two sources: first, the impact of other greenhouse gases and, second, the unresponsiveness of coal use to changing policies. Only a ban on coal instituted by the year 2000, would effectively slow the rate of temperature change and delay the date of a $2^{\circ}C$ change from a base case of 2040 to 2055. Less stringent policies such as an increase in worldwide taxes of up to 300 per cent on the cost of fossil fuels, bans on synfuels and shale oil delay the advent of a $2^{\circ}C$ rise in temperature by less than five years. Price reductions on alternative fuels have little effect in the ER model because they enhance economic performance and result in higher overall energy demand. This result is consistent with our findings. In presenting finite technical progress models, we show that an increase in energy efficiency may increase fossil fuel use, at least temporarily, because the rate of increase of efficiency tends to be lower in fossil fuel technologies than in others, and substitution is limited on a worldwide scale. The ER model examines control policies including bans on fossil fuel use, trade restrictions, and fossil fuel taxes. Because the ER model is the only major model to consider trade restrictions these are of particular interest. In one run of the

167

model a United States only energy tax is combined with a limit on energy exports. Such a policy will lower world emissions; however, it should be noted that this policy would certainly not be in the best interest of the United States in a competitive world fuel market. In a competitive world market, other suppliers would replace most US fuels taken off the market. Because the ER model includes trade, it would be useful to examine the effects of subsidies on non-fossil fuel exports such as fuel conservation equipment, nuclear energy facilities, and solar energy facilities. Our analysis of trade suggests that such subsidies may be worthwhile when international co-operation is incomplete.

Turning now to a policy analysis of the Nordhaus and Yohe model we likewise feel that the addition of feedback effects and the use of an objective function would enhance the model's results. Because the model does not link periods through state variables, a simple optimization over multiple policies would be possible. As it now stands, the model examines the effects of probability but not the particular risks of low periods of consumption. To measure the risk posed by future events we need to know both the probability of the events and how we wish to value periods of high and low consumption. Our models suggest that differences in the valuation of high and low periods of consumption may have a significant impact on present policies. Incorporation of an objective function which considered the curvature of the utility function in the NY model would allow risk to be examined in a more realistic manner.

The treatment of technical progress in this model is much more flexible than elsewhere. The approach taken, however, is historical towards the rate of technical progress; it does not explicitly link technical progress to changing economic conditions.

We also note that inclusion of feedback effects in this model might actually reduce the long-run variance in GNP, because GNP growth and the impacts of CO_2 are negatively correlated. That is, higher GNP growth raises fossil fuel use which may increase the adverse impacts of CO_2 and reduce GNP growth. In view of substantial conclusions, the NY model produced three major results: on the basis of the given evidence it first suggests that a rather high chance (25 per cent) exists for CO_2 concentrations to double before 2050. The possibility of an early doubling of CO_2 makes the incorporation of feedback effects even more important. The second is that an 11 per cent chance exists for a non-fossil backstop technology to replace fossil fuels before 2100. Another important result is the statement that the major causes of uncertainty are the ease of substitution between fossil and non-fossil fuels and the rate of technology growth. This re-emphasizes our point that structural uncertainty and sensitivity to technological change are significant factors.

Nordhaus and Yohe examined five different tax policies to control CO_2. The most stringent of these placed a 60 per cent surcharge on the price of fossil fuels. The impact of this stringent policy is to reduce the CO_2 concentrations in 2050 by 6 per cent as compared to a base case where all parameters are at their median values. It appears that part of the reason for this small change is that the response to such a tax in this model does not include changing technology and capital stocks. For example, our flexible technology model (Chapter 3) suggests that a fossil fuel tax would increase research in energy efficiency and cause a relative increase in capital purchases.

5.10 CLASSIFICATION OF EEE MODELS

During the course of our modelling approach a number of useful models were developed. The models were differentiated mainly by their production functions. In each case results were derived from the most general form of the model possible; however, the range of models developed is more clearly illustrated by specific model forms. Unless otherwise noted the objective is the maximization of discounted utility.

1. The simplest useful model with feedback effects is the single step model. This model can be used to examine international co-operation and uncertainty issues. The model assumes a critical CO_2 level exists. When below the critical level, production is a function of fossil fuel use alone; and when above the critical CO_2 level, production is zero. An equilibrium is reached in finite time, and during the transition the use of fossil fuels drops. The reaction of fossil fuel use to parameter changes is easily found in this model.

2. A slightly more complex model is the multiple step model. This model presents a very flexible and realistic method of representing CO_2 feedbacks. Economic productivity drops sharply at a finite number of CO_2 levels or steps and is constant between these steps. The use of fossil fuels drops at a constant rate between critical CO_2 levels and increases suddenly at the critical levels.

3. A Cobb-Douglas type model is a further specification of 1. This model allows the comparison of optimal policies under sudden and gradual CO_2 impacts. The harmful impacts of CO_2 occur gradually, not abruptly, with increases in CO_2. Consumption is a Cobb-Douglas function of the distances from the ultimate levels of CO_2 concentration and emissions. Exogenous exponential growth at a rate g is easily incorporated into the model. Equilibrium in this model is never reached. Changes in fossil fuel use induced by parameter changes are easily found.

4. An endogenous neutral technical progress model is proposed to facilitate the explicit treatment of investment which is not common in energy environmental models of carbon dioxide emissions. In this model the step impact of CO_2 is again used. The

170

opportunity to improve the economy through investment in knowledge or capital exists. The elasticity of production with respect to fossil fuel use, capital, and knowledge are all constant. No growth and exponential growth equilibria are both possible. The no growth equilibrium is associated with low energy sector output. The equilibria may be approached with a variety of investment patterns: invest in knowledge alone, invest in capital alone, invest in both or in neither.

5. Models of flexible technical progress make it possible to model the CO_2 problem so that the opportunities for technical change are captured in a satisfactory way. Present models do not provide a framework in which the impact of future economic conditions or policies on technical change can be examined satisfactorily. Vintage models are quite different from the models presented earlier. The objective in the first two models presented is to minimize the cost of producing a specified output. The production function in both is Leontief with a fossil fuel and a capital input. A research budget which may be used to improve fuel or capital efficiency is exogenously set in the models. From this model the major result was derived that in equilibrium technical change may be neutral when prices are stable or exponentially changing depending on the relation of energy use to output.

We feel that all the major elements in 1 to 5 should be used to build highly disaggregated, policy-oriented, flexible, large-scale models.

5.11 LESSONS LEARNED FROM MODELLING EXERCISES

By reviewing major EEE models we show that implementation of optimization and feedback could be pursued. This would be done by specifying welfare and feedback functions in these models and doing multiple runs over alternative policies. Such changes to the models would allow the examination and discussion of control policies and implications of economic growth. Inclusion of capital, especially capital in a vintage model would be much more difficult in these models and would require major efforts.

The small-scale models presented in Chapters 2-4 are useful for generating ideas, assessing large-scale models and improving modelling approaches.

As noted earlier, one important area is the examination of the possible role trade restraints and coalitions will play in the control of the international use of fossil fuels. We believe that the models presented in the previous chapters deserve more attention in building large-scale models, and improving modelling approaches. The step model used to examine the impact of uncertainty on estimates of present optimal fossil fuel use could be further refined and the subject studied in more detail. The flexible technology model with vintage capital has many features not included in other models and also deserves further work.

The CO_2 problem is well documented and imposing. It must be attacked by many disciplines and at many levels. We are limited in dealing with the problem both by our lack of understanding of the climate and the world economy.

In this overview we have not tried to forecast the future or recommend specific policies. These matters can best be dealt with through large-scale economic models.

At present most governments in the western hemisphere are limiting their actions on CO_2 to research. It appears reasonable that the research emphasis should be placed on climate, adaptation and basic economics. The study of the economic implications of climate change cannot be pushed beyond the understanding of economic growth and technical change. The estimates of the impacts of CO_2 on the economy will always be less accurate than the most accurate estimate of the size of the world economy. Our estimates of the impact of CO_2 control policies will always be less accurate than our best estimate of the future pattern of technical change in the energy industry. Large-scale models of the world economy may be used to examine the CO_2 problem, but the construction of a large-scale economic model to study the CO_2 problem alone would be wasteful.

We can identify several areas of economic research that will be particularly important

172

to CO_2. One important area is the effectiveness of research support in changing the mix of energy technologies.

Another area of research is the use of market controls, such as excise taxes, export limits, subsidies, and cartels, to reduce or increase the use of a commodity. Issues such as the effectiveness of incomplete cartels and the competitiveness of world markets in fossil and other fuels are of interest.

The CO_2 problem should be considered in determining the type of research in the energy area which government supports. From a CO_2 perspective, economic incentives for the development of coal, tar sand and oil shale are questionable. Both our models and others suggest that improvements in all non-fossil fuels do not necessarily lower future levels of CO_2. To displace fossil fuels, alternative non-fossil fuels must be highly substitutable.

Our models suggest two other factors which policy makers should be aware of when considering the economics of CO_2. As with many other environmental problems, any policies which increase the perceived discount rate will exacerbate the CO_2 problem by reducing our concern for the future. The shadow price attributed to fossil fuels by expected value economic models will be lower than the true shadow price for a risk averse society.

REFERENCES

Asako, K., "Economic Growth and Environmental Pollution under the Max-Min Principle", *Journal of Environmental Economics and Management*, 7, 1980, 157-183.

Ayres, R. U. and Miller, Steven M., "The Role of Technological Change", *Journal of Environmental Economics and Management*, 7, 1980, 353-371.

Barrett, S., "Self-enforcing International Agreements", London Business School: London, 1992.

Baumol, W. J., and Oates, W. E., *The Theory of Environmental Policy*, 2nd ed., Cambridge University Press, Cambridge, 1975.

Becker, R. A., "Intergenerational Equity: the Capital-Environmental Trade Off", *Journal of Environmental Economics and Management*, 9, 1982, 165-185.

Burniaux, J-M, Martin, J., Nicoletti, G. and J. Oliveira-Martins, "GREEN: A Multi-sector Multi-region Dynamic General Equilibrium Model for Quantifying the Costs of Curbing CO_2 Emissions", OECD Economics Department, WPNO.116, Paris, 1992.

Chiarella, C., "Optimal Depletion of a Non-renewable Resource when Technological Progress is Endogenous" in Kemp, M. C. and Long, N. V. (eds). *Exhaustible Resources, Optimality and Trade*, North Holland, Amsterdam, New York 1980.

D'Arge, R. C. and Kogiku, K. C., "Economic Growth and the Environment", *Review of Economic Studies*, 1973, 61-77.

Deshmukh, S. D. and Pliska, S. R., "Optimal Consumption of a Non-renewable Resource with Stochastic Discoveries and a Random Environment", *Review of Economic Studies*, 50, 1983, 543-554.

Deshmukh, S. D. and Pliska, S. R.. "Optimal Consumption and Exploration of Non-renewable Resources under Uncertainty", *Econometrica* 48, 1980, 177-200.

Dixit, A. K. and Norman, V., *Theory of International Trade*, Cambridge University Press, Cambridge, 1980.

European Commission (EC), ed., "Economy-Energy-Environmental Models", EC-DG XII, EUR 16712EN, Brussels, 1995.

Edmonds, J. A. and Reilly, J., "Global Energy and CO_2 to the Year 2050", *Energy Journal*, 4(3), 1983, 21-47.

Edmonds, J. A. and Reilly, J., *Global Energy: Assessing the Future*, Oxford University Press, Oxford 1985.

Forster, B. A., "Optimal Energy Use in a Polluted Environment", *Journal of Environmental Economics and Management* 7, 1980, 321-333.

Gilbert, R. J., "Optimal Depletion of an Uncertain Stock", *Review of Economic Studies*, 46, 1979, 47-57.

Gottinger, H. W., "Adoption Decisions and Diffusion: Implications for Empirical Economics", *Schweizerische Zeitschr. f. Volkswirtschaft und Statistik* (Swiss Journal for Economics and Statistics), 1991, 17-34.

Gottinger, H. W., "Greenhouse Gas Economics and Computable General Equilibrium", Journal of Policy Modelling, 1998 (forthcoming).

Hoel, M., "Emission Taxes in a Dynamic Game of CO_2 Emissions", Centre for Applied Research (SAF), University of Oslo, Mimeo, 1990.

Holly, S. and Hallet, A. H., *Optimal Control, Expectations and Uncertainty*, Cambridge University Press, Cambridge, 1989.

Loury, G. C., "The Optimum Exploration of an Unknown Reserve", *Review of Economic Studies*, 45, 1978, 621-636.

Manne, A. S., "Global 2100: Alternative Scenarios for Reducing Carbon Emissions" OECD Economics Department, WPNO.111, Paris, 1992b.

Manne, A. S. and Richels, R. G., "CO_2 Emission Limits: An Economic Cost Analysis for the USA", *Energy Journal*, 11, 1990.

Manne, A. S. and Richels, R. G., "Buying Greenhouse Insurance" in Manne, A. S. (ed.) *Global 2100: The Economic Costs of CO_2 Emission Limits*, 1991 (forthcoming).

Marchetti, C., "Primary Energy Substitution Model", *Chemical Economy and Engineering Review*, 7, 1975, 9-15.

Nordhaus, W. D., *The Efficient Use of Energy Resources*, Cowles Foundation, Yale University Press, New Haven, 1979.

Nordhaus, W. D., "Thinking about Carbon Dioxide: Theoretical and Empirical Aspects of Optimal Control Strategies", Cowles Foundation Discussion Paper, No 565, Yale University, New Haven, Connecticut, October, 1980.

Nordhaus, W. D. and Yohe, C. W., National Research Council; Carbon Dioxide Assessment Committee, Nierenberg, W. A., (ed.) *Changing Climate*, National Academy Press, Washington D. C., 1983.

Nordhaus, W. D., "Rolling the 'DICE': Optimal Transition Path for Controlling Greenhouse

Gases", *Resource and Energy Economics*, 15, 1993, 27-50.

Nordhaus, W. D., *"Managing the Global Commons: The Economics of Climate Change*, MIT Press, Cambridge, Mass, 1994.

Peck, S. C., and T. J. Teisberg, "CETA: A model for Carbon Emission Trajectory Assessment", *The Energy Journal* 13, 1992, 55-77.

Perry, A. M. et al, "Energy Supply and Demand Implications of CO_2", *Energy* 7, 1982, 991-1004.

Peterka, V., *Macrodynamics of Technological Change: Market Penetration by New Technologies*, International Institute for Applied Systems Analysis, Laxenburg, Austria, RR-77-2, November, 1977.

Rose, D. J., Miller, M. M. and Agnew, C., *Global Energy Futures and CO_2 induced Climate Change*, Report MITEL 83-015, MIT ENERGY Laboratory, Cambridge, Massachusetts, November, 1983.

Scheraga, J. D. et al, "GEMINI: An Energy-Environmental Model of the United States", A Status Report on Model Development, Workshop on Economic/Energy/ Environmental Modelling for Climate Policy Analysis, University of Tokyo/MIT Center for Energy Policy Research, October 1990.

Spinrad, B. I., *Market Substitution Models and Economic Parameters*, International Institute for Applied Systems Analysis, Laxenburg, Austria, RR-80-28, July 1980.

Ulph, A., "The Choice of Environmental Policy Instruments and Strategic International Trade", Mimeo, University of Southampton, October 1990.

CHAPTER 6

A SIMPLE ENDOGENOUS MODEL OF ECONOMIC ACTIVITY
AND CLIMATE CHANGE

6.1 INTRODUCTION

One problem that is common to the existing research on future climate change is the neglect of the forces of the market mechanism. The existing estimates on future climate change are obtained under the assumption of exogenous economic activities. Clearly, such an assumption suppresses the feedback effects to economic activities of the predicted changes in climate. For example, one of the major worries in the studies on climate change is that higher global average temperature may have severe adverse effect on world agricultural production. However, if the productivity of agriculture falls as a result of the global warming, the relative price of agricultural commodities is expected to rise, which, under certain demand conditions, will drive down the production of the manufactured goods and hence the emissions of the greenhouse gases. Taking this force into account, it is probable that existing models may have over-estimated the amount of future emissions of the greenhouse gases and their impacts on future climate.

Such criticisms on the computational models that ignore the functioning of market mechanism is not new in the economics literature. One of the major criticisms of Solow on the well-known Club of Rome Report in the early 70s is that the report fails to take into account the reaction of the built-in market mechanism (Solow, 1973). The Club of Rome Report predicts that the world economy will overshoot and then collapse some time in the middle of next century. However, once the market mechanism is incorporated into these Doomsday Models, Solow argues, the world economy will smoothly approach its natural limit, if there were indeed a limit of growth. Furthermore, as a resource becomes more and more scarce, the price mechanism will create strong incentives to reduce the consumption of the resource and to invent substitutes. Therefore, Solow argues, there is no reason to believe that such limit of growth must exist.

While the importance of market mechanism on the global warming problem is widely recognized by economists, so far there have been no studies that attempt to incorporate the market mechanism into climate modelling.[1] That is we want to explain climate change (and thus ensuing damage profiles) and exogenously, by change in economic activities (as suggested by Chichilnisky and Heal, 1993).

In order to model the dynamic interaction between economic activities and the climate system, a dynamic two-sector general equilibrium model is constructed. This Chapter attempts to answer the question, "what are the possible scenarios that would arise from the interaction between economic activities and the climate system: would the world temperature reach a steady state? or would it be increasing forever? or something else?" The answer to the above question will also provide an answer to the following questions. The greenhouse effect is supposedly caused by economic activities. Will the market mechanism itself be able to correct this problem? In other words, will the forces of the price mechanism be sufficient to stop the trend of global warming?

The model is constructed in such a way that the market has a built-in self-stabilizing mechanism that offsets rising temperature. The climate is also assumed stable. The interaction between the two stable systems, however, does not necessarily lead to a stable long-run equilibrium (Baumol and Benhabib, 1989). It is shown that the characteristics of the equilibrium time path depends critically on the planet's natural cooling tendency. The world economy and temperature will reach a steady state as long as the rate at which the plant sheds heat is not too small. If this decay is too small, however, competitive equilibrium will, under certain conditions, lead to climatic cycles or even a climatic chaos. It is shown that under certain conditions the equilibrium law of motion of temperature displays sensitive dependencies on initial conditions.

The possible existence of sensitivity in the time path of temperature casts doubts on the existing projection about future climate changes. For a dynamical system that exhibits sensitivity, a small error of measurement of the initial state may result in very large prediction errors for future dates, even if the forecaster knows very well the law of motion of the system. In reality, of course, we cannot even claim that we know the system well.

We start from the premise that emission of greenhouse gases by human activities does lead to higher global temperature. However, the main proposition of the chapter that the dynamic interaction between a stable natural system and a self-stabilizing market mechanism can, under certain conditions, lead to chaotic behaviour remains valid even if the greenhouse effect were proven to be insignificant in the future. This proposition will prove its importance over time as man's capacity to affect nature grows larger and larger as a result of technological progress.

This Chapter is organized as follows. Section 2 describes the economic and natural environment while Section 3 studies the competitive equilibrium. Section 4 analyses the properties of the law of motion of the global temperature. Sections 5 and 6 consider the equilibrium time path of temperature under different conditions based on the results obtained in Section 4. Section 7 investigates the conditions under which a climatic chaos occurs while Section 8 offers concluding remarks.

6.2   ECONOMIC AND NATURAL ENVIRONMENT

In this Section the production, consumption and climatic aspects of the model are specified. Roughly speaking, we shall consider a competitive world with no national boundaries. There are two goods in the economy, one of which is an agricultural good and the other a manufactured good. The productivity of the agricultural sector is affected by the global temperature. The manufacturing activities, on the other hand, affect the temperature level. Firms take temperature levels as given. There is no market for temperature.

The formal specification of the model follows.

Time, denoted t, is discrete and the horizon is infinite: $t \in \{0,1,...\}$.

On the *production side* of the economy, two non-storable goods are produced: an agricultural good and a manufactured good, with quantities being denoted by S_1 and S_2.

There is a fixed continuum of firms in each industry. Hence both industries are perfectly competitive. Labour is the only input of production. At each date, a representative firm in industry $i(i = 1,2)$ chooses the level of employment in the industry, ϕ_{it}.

The production technology of both goods exhibits constant returns to scale. The productivity of labour in the manufacturing sector does not depend on climate and is denoted by b. The output of the manufacturing sector at date t can then be written as $S_{2t} = b\phi_{2t}$. In the agricultural sector, however, the productivity of labour depends on one aspect of the climate, namely the global temperature. Let $a(\tau_t)$ denote the productivity coefficient of the agricultural sector, i.e., $S_{1t} = a(\tau_t)\phi_{1t}$, where τ_t is the world average temperature level in period t. It is assumed that $a(\tau_n) > 0$; $a'(\tau_t) > 0$ if $\tau_t < \bar{\tau}$ and $a'(\tau_t) < 0$ if $\tau_t > \bar{\tau}$; $a(\tau_u) = 0$; and $a''(\tau_t) < 0$. τ_n denotes the "natural" temperature level, i.e., the level at which the global temperature would stay in the absence of any manufacturing activities. $\bar{\tau}$ is some critical level of temperature for the agricultural sector, $\tau_n < \bar{\tau}$. $\tau_u(>\bar{\tau})$ is the temperature level at which the agricultural productivity equals zero. Therefore, by assumption the agricultural productivity is positive when there have been no manufacturing activities . A higher level of temperature improves the agricultural productivity as long as the temperature is below the critical value $\bar{\tau}$. As temperature levels exceed $\bar{\tau}$, however, higher temperature will reduce the productivity of the agriculture sector. The agriculture productivity eventually approaches zero as the temperature level reaches τ_u.

The *consumer side* of the economy comprises a fixed continuum of identical consumers. A representative consumer is endowed with one unit of labour endowment which

181

is supplied inelastically. He has no initial wealth. His preference over consumptions of the agricultural good and the manufactured good at date t, denoted by C_{1t} and C_{2t}, respectively, is represented by the utility function $\sum_{t=0}^{\infty} \beta^t U(C_{1t}, C_{2t})$, where $U()$ satisfies:

(A1) $U(C_1, C_2) \in C^3$ and $\lim_{C_1 \to 0} U_1(C_1, C_2) = +\infty$, $\lim_{C_2 \to 0} U_2(C_1, C_2) = +\infty$.

(A2) $U(C_1, C_2)$ is concave and homothetic.

(A3)   The elasticity of substitution between the two goods is less than unity.

We shall make more assumptions on preferences after we develop more notations in Section 4.

The law of motion of the *global average temperature* is characterised by a variation of the zero-dimensional climate system model (Dickenson, 1986[2] and Henderson-Seller and McGuffie, 1987).

$$\tau_{t+1} = (1-c)(\tau_t - \tau_n) + \tau_n + g(S_{2t}) \tag{1}$$

where $c \in (0 \; 1)$; $g(S_{2t}) \in C^2, g(0) = 0, g'(S_{2t}) > 0$ *and* $g''(S_{2t}) \le 0$.

(1) states that the manufacturing activities raises temperature[3]. When temperature is above its natural level, τ_n, the nature has the ability of absorbing a percentage of the excess greenhouse gases and cooling down the climate towards its natural level at a rate c. Since it is assumed that $\tau_n < \bar{\tau}$, starting from the point of time where no manufacturing activities had taken place in the past "some" manufacturing activities would be good for the production of the agriculture good by assumption. The world starts at a temperature level $\tau_0 \in \{\tau_n, \tau_u)$.

As a climate system model, (1) has no independent predictive value since the parameters of the model are best obtained from more detailed models. (1) is used to interpret and summarize the results of more detailed and complex models (Dickinson, 1986). Other climate system representations in energy-economy-environmental (EEE) models are possible, see Nordhaus (1993). Since the intention is to generate qualitative predictions about the outcomes of dynamic interactions between the economic system and the climate system, rather than to produce quantitative estimates of the climate system, (1) is a sufficient representation of the climate system for our purposes.

Throughout the analysis it is assumed that

(A4) $g(b) < \tau - \tau_n$,

which states that the greenhouse effect caused by one period of manufacturing activities alone

182

would not be sufficient to drive temperature from its natural level, τ_n, to $\bar{\tau}$ even if all the labour supply were devoted to manufacturing activities. (A4) can be satisfied by choosing an appropriate unit for time.

Without any loss of generality we choose the unit of temperature level in such a way that $\tau_n = 0$.

6.3   COMPETITIVE EQUILIBRIUM

We proceed to identify the elements of an intertemporal dynamic competitive equilibrium in a similar way as Boldrin (1989).

Let p_{1t} and p_{2t} denote the period t prices of the agriculture good and the manufactured good, respectively. Let w_t be the competitive wage rate prevailing in period t and R_t the competitive nominal interest rate between period $t - 1$ and t. $R_0 = 1$ by definition. The optimization problem faced by a representative consumer in this economy, in addition to the trivial decision of inelastically supplying one unit of labour, is to maximize the sum of discounted utility obtained from consuming the manufactured good and the agriculture good taking $\{p_{1t}, p_{2t}, w_t, R_t\}_{t=0}^{\infty}$ as given. The consumer's problem can then be written as:

$$\max_{\{C_{1t}, C_{2t}, A_{t+1}\}_{t=0}^{\infty}} \sum_{t=0}^{\infty} \beta^t U(C_{1t}, C_{2t}) \tag{1}$$

subject to

$$p_{1t}C_{1t} + p_{2t}C_{2t} + A_{t+1} = w_t + R_t A_t \tag{2}$$

and $A_0 = 0$. A_t is the consumer's wealth in period t.

The first-order conditions are:

$$\frac{p_{2t}}{p_{1t}} U_1(C_{1t}, C_{2t}) = U_2(C_{1t}, C_{2t}) \tag{3}$$

$$\frac{U_1(C_{1t}, C_{2t})}{p_{1t}} = \beta U_1(C_{t+1}, C_{2t+1}) \frac{R_{t+1}}{p_{1t+1}} \tag{4}$$

Define $\delta_t = \prod_{s=0}^{t} R_s$. Since each firm is infinitesimal, firms take temperature as given when choosing output level in each period. The optimization problems faced by a representative firm in the agricultural sector and, respectively, the manufacturing sector are:

$$\max_{\{\phi_{1t}\}_{t=0}^{\infty}} \sum_{t=0}^{\infty} \frac{1}{\delta_t} \left[p_{1t} a(\tau_t)^{\wedge}{}_{1t} - w_t^{\wedge}{}_{1t} \right] \tag{5}$$

184

$$\max_{\{\phi_{2t}\}_{t=0}^{\infty}} \quad \sum_{t=0}^{\infty} \frac{1}{\delta_t} \left[p_{2t} b \phi_{2t} - w_t \phi_{2t} \right] \tag{6}$$

Since there is no capital good in this model, the firms' optimization problems are in effect static. The first-order conditions are:

$$p_{1t} a(\tau_t) \geq w_t; \quad \phi_{1t} \geq 0; \quad \phi_{1t}(p_{1t} a(\tau_t) - w_t) = 0 \tag{7}$$

$$p_{2t} b \geq w_t; \quad \phi_{2t} \geq 0; \quad \phi_{2t}(p_{2t} b - w_t) = 0. \tag{8}$$

Market clearing requires that

$$C_{1t} = a(\tau_t) \phi_{1t} \tag{9}$$
$$C_{2t} = b \phi_{2t} \tag{10}$$
$$\phi_{1t} + \phi_{2t} = 1 \tag{11}$$
$$A_t = 0 \tag{12}$$

Equations (9) and (10) are goods market clearing conditions. (11) is the labour market clearing condition. Since all consumers are identical, no borrowing and lending occurs among consumers in equilibrium. This is captured by (12).

We define a competitive equilibrium in this economy as follows:

Definition 3.1. *A competitive equilibrium is a collection of sequences*
$\{P_{1t}, P_{2t}, w_t, R_t, C_{1t}, C_{2t}, \phi_{1t}, \phi_{2t}, \tau_{t+1}, A_{t+1}\}_{t=0}^{\infty}$ *and scalars* $\tau_0, A_0 = 0$ *such that:*

(1)   *given the sequence* $\{p_{1t}, p_{2t}, w_t, R_t\}_{t=0}^{\infty}$ *and scalar* A_0, $\{C_{1t}, C_{2t}, A_{t+1}\}_{t=0}^{\infty}$ *solves the consumers' problem (C).*

(2)   *given the sequence* $\{p_{1t}, p_{2t}, w_t, R_t, \tau_t\}_{t=0}^{\infty}$, $\{\phi_{1t}\}_{t=0}^{\infty}$ *solves the agricultural sector's problem (A).*

(3)   *given the sequence* $\{p_{1t}, p_{2t}, w_t, R_t\}_{t=0}^{\infty}$, $\{\phi_{2t}\}_{t=0}^{\infty}$ *solves the manufacturing sector's*

problem (M).

(4) *given* $\tau_0 \in [0, \tau_u)$, $\{\tau_{t+1}\}_{t=0}^{\infty}$ *evolves according to* $\tau_{t+1} = (1 - c)\tau_t + g(S_{2t})$. *Furthermore,* $\tau_t \in [0, \tau_u)$ *for* $t = 0, 1, 2\ldots$.

(5) $\{C_{1t}, C_{2t}, \phi_{1t}, \phi_{2t}, A_{t+1}\}_{t=0}^{\infty}$ *satisfies the market clearing conditions (3.9)-(3.12).*

Since there is no borrowing and lending in equilibrium, the consumers' optimization problem in each period is essentially static. Hence the competitive equilibrium defined above can be viewed as a sequence of one-period equilibria with temperature level in each period being pre-determined by (4) of Definition 3.1. We can thus choose one price numeraire for each period. Define the relative prices of the manufactured good, $p_t = \dfrac{p_{2t}}{p_{1t}}$ and the real rate of interest, $r_t = \dfrac{p_{1t-1}}{p_{1t}} R_t$.

Let $\psi_t = \phi_{2t}$ (11) implies that $\phi_{1t} = 1 - \psi_t$. (7) and (8) together with (3) imply that $p_t = \dfrac{a(\tau_t)}{b}$. Therefore, a competitive equilibrium is characterized by the following three conditions:

$$\frac{a(\tau_t)}{b} U_1(a(\tau_t)(1-\psi_t), b\psi_t) = U_2(a(\tau_t)(1-\psi_t), b\psi_t) \tag{13}$$

$$U_1(a(\tau_t)(1-\psi_t), b\psi_t) = \beta U_1(a(\tau_{t+1})(1-\psi_{t+1}), b\psi_{t+1})\tau_{t+1} \tag{14}$$

$$\tau_{t+1} = (1-c)\tau_t + g(S_{2t}) < \tau_u \tag{15}$$

Lemma 3.1. *Given* $\tau_t \in [0, \tau_u)$, *there is a unique interior solution to (13) for* ψ_t.
Proof: Note that, given $\tau_t \in [0, \tau_u)$, (13) is equivalent to the first-order condition to the following problem:

$$\max_{\psi_t} U(a(\tau_t)(1-\psi_t), b\psi_t)$$

which has a unique interior solution to ψ_t as a function of τ_t given the concavity assumption on $U(\cdot)$.

186

Denote this solution to (13) by $\psi^-(\tau_r)$. By the concavity of the utility function and the implicit function theorem we know that $\psi^-(\tau_r)$ is continuous and differentiable in (τ_r) .

The existence of a competitive equilibrium depends on the magnitude of c. For c sufficiently close to 0 and for some $\tau_r \in [0, \tau_u)$ the solution to (13) violates (15), in which case there will be no interior solution to r_{t+1} in (14). We will present the exact conditions under which a competitive equilibrium exists after we develop enough notations in Section 4. Here we present a "partial" existence result.

Proposition 3.2. *There exists a $\bar{c} \in (0, 1)$ such that for all $c \in [\bar{c}, 1]$ a unique competitive equilibrium exists for a given $\tau_0 \in [0, \tau_u)$.*

  Proof: Pick $\bar{c} = \dfrac{g(b)}{\tau_u}$ $(<1$ by (A4)). Since $S_{2t} \leq b$ by the labour resource constraint and (13), for all

$c \in [\bar{c}; 1]$ $\tau_{t+1} = (1 - c)\tau_t + g (S_{2t} 1) < \tau_u, \forall \tau_t \in [0, \tau_u)$.   Therefore, given

$c \in [\bar{c}, 1]$, the unique solution to (13) will also satisfy (15). From (14) one can solve for the equilibrium r_{t+1}.

  (3) and Assumption (A2) implies that in equilibrium the consumption ratio of the two goods is a function of the relative price of the two goods, i.e.,

$$\frac{C_{2t}}{C_{1t}} = \xi(p_t)$$

with $\xi(\cdot)$ being C^2. The assumptions on the utility function implies that
$\xi'(p) < 0$; $\lim_{p \to 0} \xi(p) = +\infty$; $\lim_{p \to -\infty} \xi(p) = 0$ and that

$$\frac{d[p\xi(p)]}{dp} = \xi(p)[1 + \frac{p\xi'(p)}{\xi(p)}] > 0$$

(16)

$$\lim_{p \to 0} p\xi(p) < \infty$$

(17)

$$\lim_{p \to 0} \frac{d[p\xi(p)]}{dp} = \lim_{p \to 0} \xi(p)[1 + \frac{p\xi'(p)}{\xi(p)}] = +\infty$$

(18)

In particular, (17) is derived from the fact that $p\xi(p)$ is finite for $p \in (0, \infty)$ and is increasing in p.

187

The equilibrium output of the manufactured good equals

$$S_{2t} = \frac{bp_t \xi(p_t)}{1 + p_t \xi(p_t)} \tag{19}$$

with $p_t = \dfrac{a(T_t)}{b}$. Therefore,

$$\frac{dS_2}{dp} = \frac{b \dfrac{d[p\xi(p)]}{dp}}{(1 + p\xi(p))^2} > 0 \tag{20}$$

The assumption (A3) is necessary and sufficient in determining the sign of (20). (A3) ensures that the income effect dominates the substitution effect when the productivity of agriculture (and hence the relative price) changes.

One goal of this Chapter is to answer the question whether the market mechanism can function effectively to stop the trend of global warming. We want to study the dynamic time path of global temperature and outputs in an environment where in each period the market responds to the price signals. Under (A3) and concavity assumptions on $a(\tau)$ and $g(S_2)$, this model has a built-in mechanism that serves to offset the rising temperature when τ exceeds $\bar{\tau}$; when the relative price of the manufactured good falls as a result of falling agricultural productivity, the equilibrium output of the manufactured good also falls. But is this mechanism sufficient to correct the global warming problem? In the sequel we attempt to answer this question.

6.4 PROPERTIES OF THE LAW OF MOTION OF TEMPERATURE

In this section, we shall study the properties of the law of motion of the global temperature. Strictly speaking, the results to be presented in this section are the mathematical properties of a family of maps to which the equilibrium law of motion of temperature belongs. Not all the maps in this family can be supported as the outcome of a competitive equilibrium, as we shall see later in this section.

The section starts with a list of additional assumptions on preferences and on the climate system, followed by analysis on the properties of the maps. The proofs of most results in this section are straightforward but tedious. Thus they are left to the Appendix.

Two additional assumptions on the preferences will be maintained throughout the paper:

(A5) $\lim_{p \to 0} p\xi(p) = 0$;

(A6) $\eta_{\sigma p} \equiv \dfrac{p}{\sigma} \dfrac{d\sigma}{dp} \leq (1-\sigma)$ where σ is the elasticity of substitution in consumption.

(A5) implies that $\lim_{p \to 0} S_2(p) = 0$; in other words, the equilibrium output of the manufactured good approaches zero as the relative price of the manufactured good approaches zero.

(A6) requires that the elasticity of substitution in consumption be relatively stable as price varies.

Assumptions (A5) and (A6) do not necessarily impose more restrictions on the utility functions than (A1)-(A3). In the case of a constant-elasticity of substitutions (CES) utility function, (A1)-(A3) imply (A5) and (A6). In fact, any CES utility function with elasticity of substitution less than unity satisfies all of (A1)-(A3) and (A5)-(A6).

To simplify the presentation it is assumed that

(A7) $-\tau\, D_\tau^2 g(\tau) > 1$ for all $\tau \in [\bar{\tau}, \tau_u]$.

(A7) requires that $g(S_2(\tau))$ have enough curvature in τ for τ in the interval $[\bar{\tau}, \tau_u]$. Notice that $D_\tau^2 g(\tau) < 0$ and that $D_\tau^2 g(\tau) \to -\infty$ *as* $\tau \to \tau_u$. (A7) is not a binding restriction for τ large enough. Imposition of (A7) does not change the number of possible cases in this model. It serves to restrict the occurrence of each case to one interval (as opposed to several intervals) of c values.

From (15) and (19), we can write the equilibrium time path of temperature as follows:

189

$$\tau_{t+1} = (1-c)_{\tau_t} + g[\frac{bp_t\xi(p_t)}{1+p_t\xi(p_t)}] \tag{1}$$

(1) is not well defined at $\tau_t = \tau_u$ given the existing assumptions. If we define

$$f(\tau_t;c) = \left\{ \begin{array}{ll} (1-c)\tau_t + g[S_{2t}(\tau_t)], & \textit{if } \tau_t \epsilon [0 \quad \tau_u); \\ (1-c)\tau_u, & \textit{if } \tau_t = \tau_u, \end{array} \right. \tag{2}$$

then $f(\tau_t;c)$ is continuous in τ_t in the whole interval of $[0, \tau_u]$ since

$\lim_{\tau_t \to \tau_u} \{(1-c_{\tau_t} + g[S_2(\tau_t)]\} = (1-c)\tau_u$. It is easy to verify that $f(\tau_t;c)$ is differentiable in τ_t in the interval $[0, \tau_u]$. Furthermore,

$$\frac{df}{d\tau_t} = (1-c) + \frac{dg}{d\tau_t} \left\{ \begin{array}{ll} >1-c, & \textit{if } \tau_t < \bar{\tau} \\ <1-c, & \textit{if } \tau_t > \bar{\tau}. \end{array} \right. \tag{3}$$

$$\frac{d^2f}{d\tau_t^2} = \frac{d^2g}{d\tau_t^2} < 0 \tag{4}$$

(A6) and concavity assumption on $g()$ and $a()$ are required in determining the sign of (4.4).

Notice that for $c\epsilon[0, 1]$, $f(\tau_t;c)$ constitutes a family of one-dimensional continuous maps. The following is some mathematical properties of this family of maps.

Lemma 4.1. *Assume (A1)-(A3) and (A5)-(A6). There is a unique* $\tau_s \epsilon(0, \tau_u]$ *such that* $f(\tau_s) = \tau_s$ *There is a unique* $\tau_m \epsilon(0, \tau_u)$ *such that* $f(\tau_m) \geq f(\tau)$ *for all* $\tau \epsilon[0, \tau_u]$
Proof: See Appendix

Define $\tau'_m = f(\tau_m)$. $\tau_s(c)$, $\tau_m(c)$ and $\tau'_m(c)$ are continuous functions of c. Lemma 4.1 implies that f has the shape as shown in Figure 1. Notice that since $\lim_{\tau \to \tau_u} D_\tau^2 f = -\infty$, f becomes steep very fast as τ approaches τ_u.

Recall that c is the rate at which the nature cleans up the excess amount of the greenhouse gases. As c varies from 1 (complete cleaning up) to 0 (no cleaning up), f shifts up. Analysis on how f behaves as c changes will help us to determine how the existence of an equilibrium is related to the value of c.

Lemma 4.2. *Assume (A1)-(A3) and (A5)-(A6). As c decreases, the values of τ_s, τ_m, and τ'_m*

increase.

Proof: See Appendix.

Therefore, as c decreases, both the critical point and the fixed point moves towards the right, while the maximum height of f shifts up.

Lemma 4.3. *Assume (A1)-(A7). There exists a unique $c^1 \in (0, 1)$ such that $\tau_m(c) = \tau_s(c) = \tau'_m(c)$. Furthermore, $\tau_s > \tau_m$, $\tau'_m > \tau_m$ for all $c \in [0, c^1)$; $\tau'_m < \tau_m$, $\tau_s < \tau_m$ for all $c \in (c^1, 1]$.*

Proof: See Appendix.

Lemma 4.4. *Assume (A1)-(A7). There exists a unique $c^0 \in (0, c^1)$ such that for all $c \in [0, c^0]$, $\tau'_m \geq \tau_u$ and that for all $c \in (c^0, c^1)$, $\tau'_m < \tau_u$.*

Proof: See Appendix.

With enough notations developed, we are able to give a more precise existence result than Proposition 3.2.

Proposition 4.5. *There exists a unique competitive equilibrium for $c \in (c^0, 1]$.*

Proof: By the definition of c^0, we know that the interior solution to 3(13) satisfies 3(15). Hence 3(14) can be solved for r_{t+1}.

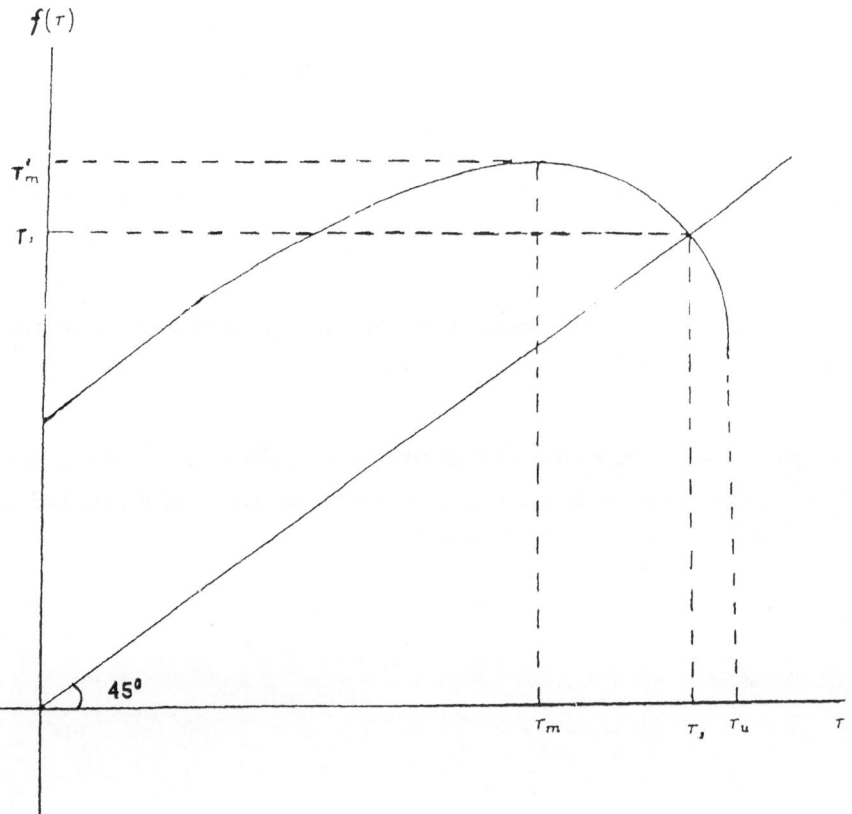

Figure 1

192

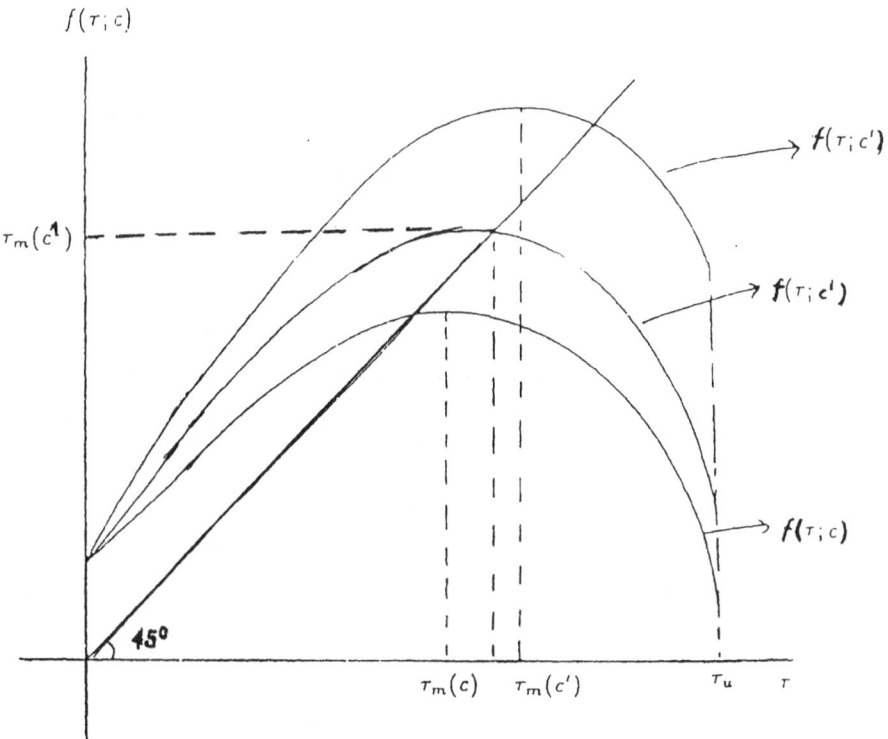

Where $c \in (c^I, 1]$ and $c' \in (c^0, c^I)$

Figure 2

To summarize the above results, c^0 and c^l $(c^0 < c^l$ divides the interval $[0, 1]$ into three segments. For c that lies in the interval $[0, c^0]$, the value of $f(\tau_l; c)$ is greater than or equal to τ_u for some $\tau_l \in [0, \tau_u]$. In this case, a competitive equilibrium in the sense of Definition 3.1 does not necessarily exist. For c that lies in the interval $(c^0, 1)$, on the other hand, $f(\tau_l;c) < \tau_u$ for all $\tau_l \in [0, \tau_u]$ and hence a competitive equilibrium exists. Furthermore, the fixed point of $f(\tau_l; c)$, τ_s, locates at the upward sloping portion of f for $c \in (c^l]$ and at the downward sloping portion of f for $c \in [c^0, c^l)$. (See Figure 2).

Remarks: In this model as the agriculture productivity falls, the equilibrium output of the manufactured good falls. As the agriculture productivity approaches 0, the equilibrium output of the manufactured good approaches 0 as well. What is going on in the case $c \in [0, c^0]$ is that, because of the assumed extremely small ability by the Earth to shed heat, the temperature rises at a much faster pace than the rise in relative price of the agriculture good in response to the greenhouse effect caused by manufacturing activities. As a result, starting from some τ_l, the global temperature may rise above τ_u before the market (prices) has time to react. In the real world, however, there is no reason to believe that the response of the market mechanism is any slower than the response of the climate system to the greenhouse effect. In fact, it appears that the contrary is true. Therefore, in the remainder of the paper we shall concentrate on the case $c \in (c^0, 1]$.

Lemma 4.6. *Assume (A1)-(A7). There exists $c_f \in [c^0, c^l)$ such that $f(\tau_m'(c)) > \tau_m(c)$ for all c in the interval (c_f, c^l).*

Proof: See Appendix.

Notice that c_f is not unique. In what follows, c_f refers to the smallest of such c only. If $c_f = c^0$, $f(\tau_m') > \tau_m$ for all c in the interval (c^0, c^l).

6.5 EQUILIBRIUM TIME PATH OF GLOBAL TEMPERATURE: CASE 1

Given the results obtained in Section 4, the behaviour of the equilibrium time path of the global temperature can be determined. As shown in the last section c^l divides the interval $(c^0, 1]$ into two sub-intervals. The time path of global temperature has different characteristics for c in the different sub-intervals. In this section, the case of c in the intervals $[c^l, 1]$ is analyzed. The analysis of the case $c \in (c^0, c^l]$ is left to Section 6 and Section 7.

The analysis in the next three sections draws heavily on definitions and results on the theory of one-dimensional dynamical system as presented in Collet and Eckmann (1980) (CE) and Grandmont (1986). Several important definitions, however, are presented below in the text.

Let $f^i(\tau)$ denote the ith iterations of a function $f(\tau)$, i.e., $f^i = of f^{i-1}$ for $i=1, 2...$, and $f^0(\tau) \equiv \tau$.

Definition 5.1. *Consider the difference equation $\tau_{t+1} = f(\tau_t)$, with f being a continuous function that maps the interval $[0, \tau_u]$ into itself. The* <u>orbit</u> *of τ is the set $\{\tau, f(\tau), f^2(\tau)...\}$. τ is said to be a* <u>periodic point</u> *of f with period k if (1) τ is a fixed point of f^k, i.e., $\tau = f^k(\tau)$; and (2) k is the smallest integer having this property, i.e., $\tau \neq f^i(\tau)$ for all $i = 1, ..., k - 1$. The corresponding* <u>periodic orbit</u> *(or* <u>cycle</u>*) is the set $\{\tau_1,...,\tau_k\}$ of all the iterates $\tau_i = f^{i-1}(\tau)$ of τ, for $i = 1, ..., k$.*

Definition 5.2. *A periodic orbit of f, $\{\tau_1,...,\tau_k\}$, is* <u>locally stable</u> *if $\mid Df^k(\tau_1) \mid < 1$. The periodic orbit is said to be* <u>weakly stable</u> *if $\mid Df^k(\tau_1) \mid \leq 1$. It is said to be* <u>superstable</u> *if $Df^k(\tau_1) = 0$.*

Note: if a periodic orbit is stable, there exists a neighbourhood U of τ_1 such that for every τ in U, $f^{kj}(\tau)$ stays in U for all $j \geq 1$ and $\lim_{j \to \infty} f^{kj}(\tau) = \tau_1$. Weak stability of a cycle allows for "one-sided" stability while superstability implies that the critical point of f, τ_m, belongs to the periodic orbit.

Let μ be a Lebesgue measure.

Definition 5.3. *f displays sensitive dependence on initial conditions if there exists a set $Y \subset [0, \tau_u]$ with $\mu(Y) > 0$ and $\varepsilon > 0$ with the following property: for every $x \in Y$ and*

195

every neighbourhood U of x, there is y $\in$ U and integer n $\geq$ 0 such that $|f^n(x) - f^n(y)|$
$> \epsilon$.

The defintion states that, if f has sensitive dependence on initial conditions, then no matter how close x and y are, they will eventually be noticeably separated under the repeated action of f. This means that the orbit of τ (under the repeated action of f) will depend in a sensitive way on the choice of the initial point τ_0.

Proposition 5.1. *Assume (A1)-(A7). If the value of c is in the interval $[c^l, 1]$, then the fixed point of f, τ_s, is the unique periodic orbit of f. Starting from $\tau_0 \in [0, \tau_u)$ the global temperature will monotonically approach τ_s, which lies in the interval $[0, \tau_m]$.*
Proof: Obvious from Lemma 4.3.

Notice that $\tau_s < \bar{\tau}$ for c close to 1 and $\tau_s > \bar{\tau}$ for c close to c^l. Hence in this case, depending on the magnitude of c, we may or may not observe falls in the world agriculture productivity. But in any event, if c is in the interval $[c^l, 1]$, the price mechanism functions to stop the trend of rising temperature and the world economy approaches a steady state.

6.6 EQUILIBRIUM TIME PATH OF GLOBAL TEMPERATURE: CASE 2

The case of $c \in (c^0, c^1)$ is more complicated than the one in Section 5. In this section, equilibrium is studied under the assumption that $c_f = c^0$. Section 7 analyses the equilibrium time path when $c_f > c^0$.

When $c \in (c^0, c^1)$, $f(\tau) < \tau_u$ for all $\tau \in [0, \tau_u]$. Hence f is a unimodal that maps $[0, \tau_m]$ into itself. Furthermore, it can be shown that for any $\tau_0 \in [0, \tau_u)$, there exists an integer N, such that $\forall n \geq N f^n(\tau_0) \in [f(\tau'_m), \tau'_m]$. .

Proposition 6.1. *Assume (A1)-(A7). If $c \in (c^0, c^1)$ and $c_f = c^0$, then for any $\tau_0 \in [0, \tau_u)$, τ_t converges to either a periodic orbit of order 1 (i.e., a steady state) or a periodic orbit of order 2 that lies in the interval $[f(\tau'_m), \tau'_m]$.*

Proof: See Appendix.

Proposition 6.2. *Assume (A1)-(A7). If $c \in (c^0, c^1)$ and $c_f = c^0$, then $[f(\tau'_m), \tau'_m]$ contains a weakly stable periodic orbit of f.*

Proof: See Appendix.

Proposition 6.1 states that for any initial temperature in the interval $[0, \tau_u)$, $\{\tau_t\}_{t=0}^{\infty}$ will converge to a periodic orbit which may or may not be stable. Proposition 6.2 states that at least for some $\tau_0 \in [0, \tau_u)$ the global temperature will converge to a stable periodic orbit.

Stronger results can be obtained by assuming a negative Schwarzian derivative, i.e.,

(A8) f is C^3 and satisfies

$$S[f(\tau)] = \frac{D_\tau^3 f}{D_\tau f} - \frac{3}{2} [\frac{D_\tau^2 f}{D_\tau f}]^2 < 0$$

wherever $D_\tau f \neq 0$.

Proposition 6.3. *Assume (A1)-(A8). If $c \in (c^0, c^1)$ and $c_f = c^0$, then there exists a unique weakly stable periodic orbit P which lies in the interval $[f(\tau'_m), \tau'_m]$ and which attracts the critical point τ_m. The period of P is 1 or 2.*

Proof: Follows from Propositions 6.1 and 6.2 and Grandmont's (1986) Proposition 3.

Proposition 6.4. *Assume (A1)-(A-8), $c \in (c^0, c^1)$ and $c_f = c^0$. There exist at most two periodic orbits, one of which is weakly stable. Furthermore, (a) if $-D_\tau g(\tau_s) \leq 2 - c$, then the fixed point, τ_s, is the unique periodic orbit; (b) if $-D_\tau g(\tau_s) > 2 - c$, then, in addition to the fixed point, there exists a (unique) two-period cycle P. P is weakly stable.*

Proof: See Appendix.

Proposition 6.5. *Assume (A1)-(A8), $c \in (c^0, c^1)$, and $c_f = c^0$. Let E be the set of τ_0 such that $f^n(\tau)$ does not converge to P. Then $\mu(E) = 0$. In other words, $f^n(\tau)$ converges to P for almost all $\tau_0 \in [0, \tau_w)$.*

Proof: Follows from Proposition 6.3 and CE II.5.7.

Proposition 6.6. *Assume (A1)-(A8). If $c \in (c^0, c^1)$, and $c_f = c^0$, then f is not sensitive to initial conditions.*

Proof: Follows from Proposition 6.4 and CE II.7.1.

The results in this section have been obtained under the assumption $c_f = c^0$. The following proposition offers a necessary and sufficient condition under which $c_f = c^0$.

Proposition 6.7. *Assume (A1)-(A7). $c_f = c^0$ if and only if*

$$- \frac{g(\tau'_m) - c\tau'_m}{g(\tau_m) - c\tau_m} < 1 \tag{1}$$

for all $c \in (c^0, c^1)$.

Proof: See Appendix.

Recall that $g(\tau_t)$ represents the increase in next period's temperature caused by manufacturing activities while $c\tau_t$ is the reduction in next period's temperature by nature when the current temperature is at τ_t. Suppose that temperature reaches τ_m in period T. Then temperature will reach its maximum level, τ'_m, in period $T+1$ and will fall in period $T+2$. (1) states that the magnitude of decrease in temperature between period $T+1$ and $T+2$ must be smaller than the magnitude of increase in temperature between T and $T+1$. Therefore, Proposition 6.7 implies that the fluctuation in the size of manufacturing activities must be relatively small in order to obtain simple dynamics (steady state or two-period cycle) for temperature.

Proposition 6.7 can be restated in terms of restrictions on the greenhouse effect term

$g(\tau)$.

Corollary 6.8. *Assume (A1)-(A7).* $c_f = c^0$ *if and only if* $g(\tau)$ *satisfies*

$$- \frac{g[g(\tau)(a+\eta_{g\tau})] - [1+D_\tau g(\tau)]g(\tau)(1+\eta_{g\tau})}{g(\tau) - (1+D_\tau g(\tau))\tau} < 1 \tag{2}$$

for all $\tau \in (\tau_m(c^l), \tau_m(c^0))$ *with* $\eta_{g\tau} = -\dfrac{\tau D_\tau g(\tau)}{g(\tau)}.$

   Proof: See Appendix.

Remarks on Assumption (A8): The Schwarzian derivative plays an important role in the theory of one-dimensional dynamical systems. The implications of this condition in the context of this model, however, are not easy to derive since it involves third-order derivatives whose expressions are complicated. The remainder of this section attempts to relate the Schwarzian derivative to the primitives of the model by offering a set of necessary conditions for $S[f(\tau)] < 0$ and some examples where these conditions may (or may not) be satisfied.

Proposition 6.9. *Assume (A1)-(A6). At least one of the following must be true for* $S[f(\tau)] < 0$: *either*   *(a)* $S[g)S_2)] < 0$; *or (b)* $S[p\xi(p)] < 0$; *or (c)* $S[a(\tau)] < 0$;   *or* *(d)* $D_\tau^3 < 0$. *Furthermore, conditions (a)-(d) holding simultaneously is sufficient for* $S[f(\tau)] < 0$.

   Proof: $S[f(\tau)] < 0$. *if and only if*

$$(D_\tau^3 f)(D_\tau f) - \frac{3}{2}(D_\tau^2 f)^2 < 0. \tag{3}$$

Calculation gives

$$(D_\tau^3 f)(D_\tau f) - \frac{3}{2}(D_\tau^2 f)^2$$

$$= [(D_{s_2}^3 g)(D_{s_2} g) - \frac{3}{2}(D_{s_2}^2)](D_p S^2)^4 (D_\tau p)^4$$

$$+ (D_p^3(p\xi(p))D_p(p\xi(p)) - \frac{3}{2}(D_\tau^2 a)^2(p\xi(p)))^2](D_{(p\xi(p))}S_2)^2(D_{s_2} g)^2(D_\tau p)^4. \tag{4}$$

$$+ (\frac{1}{b})^2[(D_\tau^3 a)(D_\tau a) - \frac{3}{2}(D_\tau^2 a)^2](D_{s_2} g)^2(D_p S_2)^2$$

$$+ (1-c)D_\tau^3 f$$

199

At least one of the four terms in the right-hand side of (4) must be negative for $S[f(\tau)] < 0$ to hold. Furthermore, $S[f(\tau)] < 0$, if all four terms are negative.

Examples can be found where specific functional forms of $g(S_2)$, $p\xi(p)$ and $a(\tau)$ have the property of a negative Schwarzian derivative. It can be verified that $S[g(S_2)] < 0$ *if* $g(S_2)$ is a third-degree polynomial in S_2 with appropriately chosen parameter values. It is also easy to check that $a(\tau)$ has a negative Schwarzian derivative if $a(\tau)$ is a quadratic function in τ. The function $p\xi(p)$ derived from a CES utility function satisfies the condition of negative Schwarzian derivative if the elasticity of substitution is greater than 2. This last condition, however, contradicts Assumption (A3).

Consider an example where the functional forms of $g(S_2)$ and $a(\tau)$ are as described above and where the utility function is CES with the elasticity of substitution less than unity. It can be shown that $D_\tau^3 f < 0$ for $\tau > \bar{\tau}$ and $D_\tau^3 f > 0$ for $\tau < \bar{\tau}$ Since $g(S_2)$ and $a(\tau)$ have negative Schwarzian derivatives, $S[f(\tau)] < 0$ can be satisfied as long as the first and the third term dominate the second and the fourth terms in the right-hand side of (4).

6.7   CHAOTIC CLIMATE DYNAMICS

In this section, the possibility of the emergence of chaotic climate is explored. In Section 6, it has been shown that under (A1)-(A7) and $c_f = c^0$, the longest possible periodic orbits are of period two. If $c_f > c^0$, however, it is possible that competitive equilibrium leads to a climatic chaos, as we shall see in this section.

Define τ_k as the smaller of τ that satisfies $f(\tau) = \tau_m$. $(\tau_k < \tau_m)$. A climatic chaos may occur if there exists $c_e \in (c^0, c_f)$ such that $f(\tau'_m(c_e)) < \tau_k (c_e)$. In the following analysis, it is assumed that (A9) c_e exists.

It is straightforward to verify that under (A1)-(A7) and (A9), $f(\tau;c)$ with $c \in [c_e, c^l]$ constitutes a full family of C^l -unimodal maps. The following results are direct applications of the results in Grandmont (1986).

Proposition 7.1.   *Assume (A1)-(A7) and (A9).   Consider* $c \in [c_e, c^l]$.

(a)   *Given an arbitrary* $k \geq 2$, *the set of c for which the map* $f(\cdot;\, c]$, *has a superstable cycle of period k is closed and non-empty. Given such a c, there is an open interval around c such that* $f(\cdot;c)$ *has a stable cycle of period k for all c' in this interval.*

(b)   *Let* c_j^* *be the largest value of the parameters c for which a superstable cycle of period* 2^j *obtains for* $j \geq 1$. *Then the sequence* c_j^* *decreases with j and converges to some value* $c_\infty^* > c_e$ *as j tends to* $+\infty$. *For each c in* $(c_\infty^*, c^l]$, *all cycles of the map* $f(\cdot;c')$ *have a period that is a power of 2 or are fixed points. The critical point* $\tau_m(c)$ *of* $f(\cdot;c')$ *is attracted to one of these.*

(c)   *If superstable cycles of periods* 2^j *and* $2^{j'}$ *with* $j' > j+1$ *occur respectively for the values c and c ' in* $(C_\infty^*, c^l]$, *then a superstable cycle of period* 2^i *with* $j' > i > j$ *must appear for some value in the open interval determined by c and c'.*
  Proof: See the proof of Grandmont's (1986) Theorem 7.

Proposition 7.2.   *Assume (A1)-(A9).*

(a)   *For any c in* $[c_\infty^*, c^l]$, *the map* $f(\cdot \; ; c)$ *has a (unique) weakly stable periodic orbit.*

(b)   *There is an uncountable set of values of c in* $[c_e, c_\infty^*)$ *for which* $f(\cdot \; ; c)$ *has no stable periodic orbit.*

  Proof: See Theorem 9 in Grandmont (1986).

Proposition 7.3. *Assume (A1)-(A9).*

(a)   *For any c in $[c_{\infty}^{*}, c^{1}]$, the map $f(\cdot; c)$ has no sensitivity to initial conditions*

(b)   *There is an uncountable set of values of c in $[c_e, c_{\infty}^{*})$ for which $f(\cdot;c)$ has sensitivity to initial conditions.*

Proof: (a) follows from Proposition 7.2(a) and CE II.7.1.   (b) follows from Grandmont (1986) Remark 1.

Since a necessary condition for $f(\cdot;c)$ to display sensitivity is that $f(\cdot;c)$ has no stable periodic orbit, the set of c values defined in Proposition 7.2(b) contains the set of c values in Proposition 7.3(b).

Remarks: In this section, it is demonstrated that under (A1)-(A9) c_{∞}^{*} divides the interval $(c^0, c^1]$ into two sub-intervals.   If $c \in (c_{\infty}^{*}, c^1]$ the world economy and temperature will converge to a unique steady state or a unique cycle of period 2^j ($j = 1, 2,$ or $3...$) for almost all initial conditions $\tau_0 \in [0, \tau_u)$.   However, chaos and sensitive dependence on initial conditions may be observed if the value of c falls in the interval (c^0, c_{∞}^{*}).   In particular, the interval (c_e, c_{∞}^{*}) contains an uncountable set of such c values.[4]

The possible existence of sensitivity in the time path of temperature casts doubts on the existing projection about future climate changes.   For a dynamical system that exhibits sensitivity, a small error of measurement of the initial state may result in very large prediction errors for future dates, even if the forecaster knows very well the law of motion of the system.   In the reality, of course, we cannot even claim that we know the system well.

It is worth pointing out that chaotic climate may occur even if the climate system itself is stable.   As we can see from 2(1), in the absence of any manufacturing activities the climate system has a globally stable steady state τ_n.   It is the human activities that generated the possibility of chaotic climate.   Hence, knowing the climate system alone is not enough for projecting future climate.

Intuitively speaking, the climatic cycles and chaos are a consequence of the dynamic interaction between the stable climate system and the self-stabilizing economic system.   To illustrate this, consider the extreme cases where only one of these two systems controls the law of motion of temperature.   First, we consider the case $c = 1$ (no accumulation of the greenhouse gases).   The equilibrium law of motion of temperature becomes $\tau_{t+1} = g(S_{2t}(\tau_t))$.   Temperature in period t is completely determined by the size of the manufacturing activities in the previous period.   This is the case where the law of motion of temperature is controlled

202

by the market mechanism. In this case, temperature converges to a steady state (Proposition 5.1) because the market mechanism is self-stabilizing. Second, consider the case where the market system does not respond to the temperature changes. To be more specific, suppose in each period there is a fixed amount of the greenhouse gases emitted into the atmosphere that will raise temperature by $\bar{g}$ units. The law of motion of temperature in this case is $\tau_{t+1} = (1-c)\,\tau_t + \bar{g}$. In this case, temperature again converges to a steady state, $\dfrac{\bar{g}}{c}$, because the climate system is stable. Therefore, it is the interaction between the two stable systems that leads to the possibility of climatic cycles or chaos.

In this model, the climate system has the tendency to move temperature towards its steady state level. The market mechanism, at the same time, also works to reduce the output of the manufactured good when temperature exceeds $\bar{\tau}$. The combined forces of the two systems, under certain conditions, increase the amplitude of fluctuation. As a result, cycles or even chaos occur. Hence, in this model the self-stabilizing market mechanism does not always serve the function of stabilizing temperature. In fact, under certain conditions, the market mechanism works to "sabotage" the functioning of the climate system and to destabilize temperature.

Finally, the following result provides a sufficient condition under which (A9) is true.

Proposition 7.4. *Assume (A1)-(A7) and $c_f > c^0$. (A9) holds if there exists $c \in (c^0, c_f)$ such that for all $v \in [f(\tau_m'),\ \tau_m]$,*

$$-D_\tau^2 g(v)\ \frac{2(\tau_m' - \tau_m)}{[f(\tau_m' - \tau_m]^2}$$
(1)

Proof: See Appendix.

203

6.8 CONCLUDING REMARKS

This Chapter offers an alternative approach to the greenhouse effect modelling by incorporating endogenous economic activities into a simple climate system model. It was demonstrated that even though the climate system itself is stable and the market mechanism works to offset rising temperatures, global temperature will still display a cyclical or even chaotic time path under certain conditions. Therefore, the answer to the question posed in the title is: yes, under certain conditions economic activities will indeed lead to a climatic chaos.

We point out and prove the possibility that the interaction between a stable natural system and a self-stabilizing market mechanism can lead to cyclical or even chaotic behaviour. A built-in self-stabilizing market mechanism will not always serve the function of stabilization. Under certain conditions, it may increase the amplitude of fluctuations and have the effect of de-stabilization, as demonstrated in this paper. Therefore, by incorporating a self-stabilizing market mechanism this model yields a result that runs counter to Solow's (1973) conjecture that the market mechanism will have the effects of smoothing the time path of the world economy. Therefore, one should not take Solow's arguments as universally applicable to all other environmental issues such as the global warming.

One implication of the model is that the informal approach proposed by Lave (1982) may not be sufficient to generate reliable predictions. In this model, if one follows Lave's proposed approach by letting an economist and a climatologist independently compute their "internally consistent scenarios" using models of their own disciplines and then interact with each other, what will be generated from this process is that both will arrive at the conclusion that the world will reach a steady state. The possibility of cycles and chaos will advertently be neglected in this process. Therefore, a formal, unified general equilibrium model is needed in order to generate reliable predictions.

In this model, it has been assumed that there is no capital good. The rationale for this assumption is that climate change is a very slow process. The atmospheric residence time of CO_2 is hundreds of years. Climate change is measured in terms of time scales that are in the magnitude of hundreds of years as well. Most capital goods depreciate completely within such a time period. Therefore, the issue of investment and capital accumulation can be considered irrelevant in a model with such a long period.[5]

In this Chapter, no attempt is made to calibrate the model. The main goals are to generate qualitative predictions about the time path of temperature and at the same time to offer an example to illustrate the proposition that the dynamic interaction between a natural system and a self-stabilizing market mechanism can lead to cyclical or even chaotic behaviour. The theoretical model seems to be sufficient to accomplish these objectives. Nevertheless, the main propositions would be more forceful if one could calibrate the model using parameter values that are within the range established by empirical studies and derive cyclical and chaotic behaviour from the calibrated model. Thus calibration of the model is an interesting exercise that warrants future work.

Finally, it should be emphasized that while climatic cycles and chaos are possible, they are by no means inevitable. First, climatic cycles and chaos will occur only under certain conditions. It was shown that the world will converge to a steady state under other conditions. Second, and more importantly, the results in this Chapter are obtained under the implicit assumption of stationary social and economic structure. While it is extremely difficult to predict with certainty the changes in social and economic structure in the next one hundred years, it is conceivable that the world will undergo tremendous social and economic changes as the decades go by. (To see this possibility, one needs only to think of how much the world has changed in the past one hundred years.) These social and economic changes will undoubtedly have profound influences on the way in which the human race will interact with the nature and the way in which the climate changes will affect the well-being of the human race. (For a more detailed discussion on these issues, see Schelling, 1983.)

APPENDIX

This appendix relies on definitions and results in the theory of one-dimensional dynamic systems, Collet and Eckmann (CE) (1980) and Grandmont (1986), see also Majumdar (1994).

A1   *Proof of Lemma 4.1*: Since $f(0) = g[S_2(0)] > 0$ and $f(\tau_u) = (1-c)\,\tau_u \leq \tau_u$, there exists $\tau_s \in [0, \tau_u]$ such that $\tau_s = f(\tau_s)$ by the intermediate value theorem. The concavity of f implies that the fixed point is unique.

From 3(20)

$$\lim_{p \to 0} \frac{dS_2}{dp} = +\infty \tag{A1.1}$$

which implies that

$$\lim_{\tau \to \tau_u} \frac{df}{d\tau} = -\infty \tag{A1.2}$$

Since $D_\tau f(0) > 0$, by the continuity of $D_\tau f(\tau)$ there exists a unique critical point τ_m, for f. Since f is concave, $f(\tau_m)$ is a maximum.

Proof of Lemma 4.2:   The concavity of f implies that $f(\tau) < f(\tau_s) + f'(\tau_s)(\tau - \tau_s)$. Since $f(\tau) > \tau$ for $\tau < \tau_s$, one can prove that $D_\tau f(\tau_s) < 1$, which implies $D_\tau g(\tau_s) < c$. τ_s satisfies $c\tau_s = g[S_2(\tau_s)]$. Total differentiation reveals:

$$\frac{d\tau_s}{dc} = \frac{\tau_s}{(D_\tau g(\tau_s) - c)} < 0 \tag{A1.3}$$

The critical point of f satisfies $(1-c) = -D_\tau g(\tau_m)$. Total differentiation gives:

$$\frac{d\tau_m}{dc} = \left(\frac{d^2 g}{d\tau^2}\Big|_{\tau_m}\right)^{-1} < 0 \tag{A1.4}$$

Since $\tau'_m = (1-c)\tau_m + g(\tau_m)$, the envelope theorem implies:

$$\frac{d\tau'_m}{dc} = -\tau_m < 0 \tag{A1.5}$$

206

Proof of Lemma 4.3: When $c = 1$, $f(\tau) = g(\tau)$ and $\tau_m = \tau$. Since $g(S_2) \leq g(b)$, (A4) implies that at $c = 1$, $\tau_s < \tau_m$. When $c = 0$, $\tau_s = \tau_u$. $D_\tau g(\tau_m(0)) = -1$ implies that $\tau_m(0)$ is in the interior of $[0, \tau_u)$. Hence $\tau_s > \tau_m$ at $c = 0$.

Since $\tau_s < \tau_m$ at $c = 1$ and $\tau_s > \tau_m$ at $c = 0$, the continuity of $\tau_m(c)$ and $\tau_s(c)$ implies that there exists at least one $c' \in (0, 1)$ such that $\tau_m(c) = \tau_s(c) = \tau'_m(c)$. Such c is unique because

$$\frac{d(\tau_s - \tau_m)}{dc}\bigg|_{\tau_s = \tau_m} = -\tau_m - \frac{1}{D_\tau^2 g(\tau_m)} < 0. \tag{A1.6}$$

Otherwise, $\tau_s(c) - \tau_m(c)$ will be upward sloping for at least one of such c.

Therefore, by the continuity of τ_m and τ_s in c, $\tau_s > \tau_m$ for $c \in (0, c')$ and $\tau_s < \tau_m$ for $c \in (c', 1]$.

Notice that $\tau_s > \tau_m$ implies that $f(\tau_m) > \tau_m$, i.e, $\tau'_m > \tau_m$. Similarly, $\tau_s < \tau_m$ implies that $\tau'_m < \tau_m$.

Proof of Lemma 4.4: $\tau_s(0) \neq \tau_m(0)$. Thus, $\tau'_m(0) > \tau_s(0) = \tau_u$ by the definition of τ_m. At c', $\tau'_m = \tau_m < \tau_u$. (A1.5) implies that there exists a unique $c^0 \in (0 \; c')$ such that $\tau'_m = \tau_u$.

Proof of Lemma 4.6: Comparative statics gives:

$$\frac{d(f(\tau'_m) - \tau_m)}{dc} = -f'(\tau'_m)\tau_m - \tau'_m - \frac{1}{D_\tau^2 g(\tau_m)} \tag{A1.7}$$

Since at $c = c'$, $\tau_m = \tau'_m = \tau_s$ and $-\tau_m D_\tau^2 g(\tau_m) > 1$, then $\frac{d(f(\tau'_m) - \tau_m)}{dc}\big|_{c=c'} < 0$. Hence there exists $c_f \in [c^0, c')$ such that $f(\tau'_m) > \tau_m \; \forall c \in (c_f, c')$.

A2 *Proof of Proposition 6.1:* $c_f = c^0$ implies that $f(\tau'_m) > \tau_m$ for $c \in (c^0, c')$. Hence the extended itinerary of τ_0, $I_E(\tau_0) = AR^\infty$, is eventually periodic. Since the sequence of $\{R\}$ contains an odd number of R, the conclusion follows by applying CE II.3.1 and II.3.2.

Proof of Proposition 6.2: Define two identical open sets, $K = W = (f(\tau'_m), \tau'_m)$. Since for $\tau_t \in (f(\tau'_m), (\tau'_m), f(\tau_t) \in (f(\tau'_m), \tau'_m)$,

$$f^n(K) \subset (f(\tau'_m), \tau'_m) = K = W$$

207

$\forall h \geq 1$. Thus $f|_w$ has a sink. Thus, $W = [f(\tau'_m), \tau'_m]$ contains a weakly stable period orbit by CE II.5.1.

Proof of Proposition 6.4: Proposition 6.1 states that all possible periodic orbits are of period one or two. Hence we can concentrate on the graph of $\varphi(\tau) \equiv f^2(\tau)$.

Define τ_k as the smaller of τ that satisfies $f(\tau) = \tau_m$. $(\tau_k < \tau_m)$. Then $\varphi(\tau)$ is increasing in the interval $[0, \tau_k)$ and decreasing in (τ_k, τ_m) with τ_k being a local maximum of $\varphi(\tau)$. Similarly $\varphi(\tau)$ is increasing in the interval $[\tau_m, \tau'_m]$ with τ_m being the local minimum of $\varphi(\tau)$. Figure A1 displays the shape of $\varphi(\tau)$. $\varphi(\tau)$ has at least one fixed point, namely τ_s.

Suppose f has more than two periodic orbits. Then $\varphi(\tau)$ must have at least three fixed points in the interval $(\tau_m, \tau_s]$. Let τ^1, τ^2 and τ^3 with $\tau_m < \tau^1 < \tau^2 < \tau^3 \leq \tau_s$ be three consecutive fixed points of φ. As proven in CE p.98, $S(f) < 0$ implies that $D\varphi(\tau^2) > 1$. Since $D\varphi(\tau) > 0$ for $\tau \in [\tau_m, \tau_s]$, it must be true that $0 < D\varphi(\tau^1) < 1$ and that $0 < D\varphi(\tau^3) \leq 1$, which implies that $\varphi(\tau)$ has more than one weakly stable periodic orbits; a contradiction to Proposition 6.3.

Proof of (a): Let τ^1 denote the smallest fixed point of φ. $\tau^1 > \tau_m$ since $\varphi(\tau_m) = f(\tau'_m) > \tau_m$. τ^1 belongs to a weakly stable periodic orbit because $0 < D\varphi(\tau^1) \leq 1$.

$-D_r g(\tau_s) \leq 2-c$ implies that $-D_r f(\tau_s) \leq 1$, i.e, τ_s is weakly stable. Then $\tau_s = \tau^1$ by Proposition 6.3. It is impossible that $\varphi(\tau)$ has other fixed point by the definition of τ^1.

Proof of (b): $-D_r g(\tau_s) > 2-c$ implies that $-D_r f(\tau_s) > 1$, i.e, τ_s is unstable. Hence τ_l must belong to a weakly stable two-period cycle.

Proof of Proposition 6.7:

$$
\begin{aligned}
f(\tau'_m) - \tau_m &= (1-c)\tau'_m + g(\tau'_m) - \tau_m \\
&= -c\tau'_m + g(\tau'_m) + \tau'_m - \tau_m \\
&= [g(\tau'_m) - c\tau'_m] + [g(\tau_m) - c\tau_m]
\end{aligned}
\tag{A2.1}
$$

The result follows from (A2.1) by utilizing the fact that $g(\tau_m) > c\tau_m$ for $c \in (c^0, c^1)$.

Proof of Corollary 6.8: For $c \in (c^0, c^1)$, $\tau_m \in (\tau_m(c^1), \tau_m(c^0))$.
Since $(1-c) + g'(\tau_m) = 0$,

208

$$c = 1 + g'(\tau_m) \qquad (A2.2)$$

Therefore,

$$\tau_m' = -g'(\tau_m)\tau_m + g(\tau_m)$$

$$(A2.3)$$

$$= g(\tau_m)[1 + \eta_{gr}(\tau_m)]$$

(6.2) is obtained by substituting (A2.2) for c and (A2.3) for τ_m' into 6(1).

Proof of Proposition 7.4: $f(\tau_m) \equiv \tau_m' > \tau_m$ and $f(\tau_m') < \tau_m$ since $c \in (c^0, c_f)$. Take Taylor expansion of $f[f(\tau_m')]$ around τ_m and notice that $D_{\tau}^2 f(\tau) = D_{\tau}^2 g(\tau)$:

$$f^2(\tau_m') - \tau_m = f(\tau_m) - \tau_m + \frac{D_{\tau}^2 g(v)}{2} [f(\tau_m') - \tau_m]^2 < 0 \qquad (A2.4)$$

for $v \in [f(\tau_m') < \tau_m]$, under the condition given in the propostion.

Therefore, $f^2(\tau_m') < \tau_m$, i.e, $f(\tau_m') < \tau_k$, given this particular c.

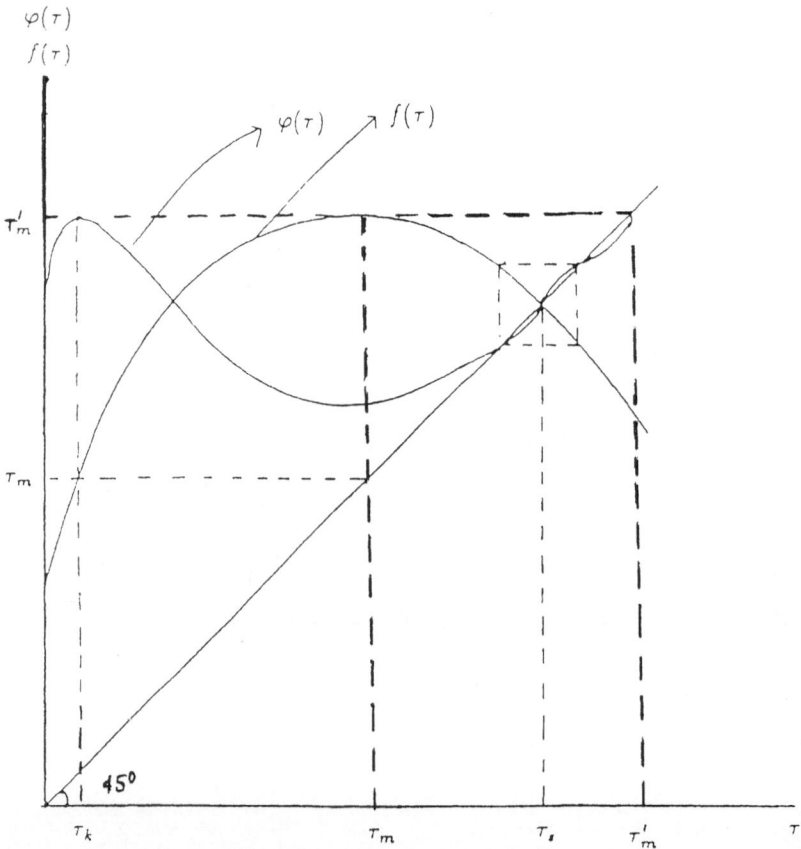

Figure A1

Footnotes

1. Most of the discussions and research conducted by economists are centred on the policy issues and the costs and benefits associated with combating the global warming. (See, for example, Whalley and Wigle, 1991; Manne and Richels, 1990; Nordhaus, 1990, 1990.

2. The original model in Dickinson (1986) is in the form of a differential equation which can be written as (in our notions):
$$\frac{\delta \Delta \tau_t}{\delta t} + \frac{c}{\Delta t} \Delta \tau_t = g(S_{2t})$$
with $\Delta \tau_t = \tau_t - \tau_n$ and Δt being the length of a period. c is equal to Δt times the ratio temperature feedback parameter to the system heat capacity parameter.

3. While deforestation in order to expand agriculture activities itself also generates additional CO_2, the net emission from this source is insufficient to bring about a significant change of climate (Bolin, Jäger and Döös, 1986, p.7). Hence we assume here that the agricultural sector does not generate the greenhouse effect.

4. Nothing is said about the case $c \in (c^0, c_e)$. This is because the analysis on $c \in [c_e, c']$ has generated all possible scenarios. An analysis on the case $c \in (c^0, c_e)$ can be conducted in the same way but will not generate any new scenarios.

5. Nordhaus (1990) discusses the difficulties in choosing appropriate time discount factors for computing the cost of future climate change. A rate close to the return on capital in most countries, Nordhaus explains, would imply that one should forget about climate change for a few decades. Hence a much lower time discount rate is required in order to derive plausible estimates of future climate damages. As a result, one is faced with the dilemma of a low time discount rate on climate change and a high return on capital (Nordhaus, 1990, p.205). One explanation for such a dilemma is that the rate of return on capital is irrelevant to the calculation of the future climate damages because capital goods are "perishable" when considered in terms of the time scale of the greenhouse effect models.

REFERENCES

Baumol, W. J. and J. Benhabib, "Chaos: Significance, Mechanism, and Economic Applications", Journal of Economic Perspectives 3(1), 1989, 77-105.

Boldrin, M., "Paths of Optimal Accumulation in Two-Sector Models", Chapter 10 in Barnett, W.A. (eds.), Economic Complexity, Cambridge University Press: Cambridge, 1989.

Bolin, B., J. Jäger, B.R. Döös (1986), "The Greenhouse Effect, Climatic Change, and Ecosystems, A Synthesis of Present Knowledge", in Bolin, et. al., eds., *The Greenhouse Effect, Climatic Change and Ecosystems*, John Wiley & Sons.

Bullard, J. and A. Butler, (1993), "Non-linearity and Chaos in Economic Models: Implications for Policy Decisions", *Economic Journal* 103, pp.849-867.

Chichilnisky, G and G. Heal (1993), "Global Environmental Risks", *Journal of Economic Perspectives* 7, pp.65-86.

Collet, P. and J.P. Eckmann (1980), *Iterated Maps on the Interval as Dynamical Systems*. (Progress in Physics: 1), Birkhauser.

Dickinson, R.E. (1986), "How Will Climate Change? The Climate System and Modelling of Future Climate," in Bolin, et. al., eds., *The Greenhouse Effect, Climate Change and Ecosystems*. John Wiley & Sons.

Grandmont, J. (1986), "Periodic and Aperiodic Behaviour in Discrete, One-dimensional, Dynamical Systems." In Hildenbrand, W. and A. Mas-Collel, eds., *Contributions to Mathematical Economics*, North-Holland, New York.

Henderson-Seller, A. and K. McGuffie (1987), *Climate Modelling Primes*, John Wiley & Sons, Chichester, New York.

IPCC (1990), Intergovernment Panel on Climate Change, *Climate Change*, J.T. Houghton et al., eds., Cambridge University Press: Cambridge.

Lave, L. (1982), "The Greenhouse Effect: The Socioeconomic Fallout" in D. Abrahamson and P. Ciborowski, eds., *The Greenhouse Effect*, Minneapolis: University of Minnesota Press.

Majumdar, M. (ed.), (1994), "*Chaotic Dynamical Systems*, Economic Theory (Special Issue), 4(5), pp. 641-790.

Manne, A.S. and R.G. Richels (1990), "CO_2 Emission Limits, An Economic Cost Analysis for the USA," *Energy Journal*, 11:2:51-74.

Nordhaus, W.D. (1990), "Global Warming: Slowing the Greenhouse Express", in Henry J. Aaron, ed., *Setting National Priorities: Policy for the Nineties*, The Brookings Institution, Washington D.C.

Nordhaus, W.D. (1991), "To Slow or Not to Slow: The Economics of the Greenhouse Effect," *The Economic Journal* 101, pp.920-937.

Nordhaus, W.D. (1993), "Rolling the Dice: An Optimal Transition Path for Controlling Greenhouse Gases', *Resource and Energy Economics* 15, 1993, pp. 27-50.

Schelling, T.C. (1983), "Climate Change: Implications for Welfare and Policy", in National Research Council, National Academy Press, pp. 449-482.

Solow, R. (1973), "Is the End of the World at Hand?" in *The Economic Growth Controversy*, International Arts and Sciences Press Inc., White Plains, New York.

Whalley, J. and R. Wigle (1991), "Cutting CO_2 Emission: The Effects of an Alternative Policy Approach", *Energy Journal* 12(1), pp. 109-124.

CHAPTER 7

GLOBAL POLLUTION AND
ECONOMIC GROWTH

7.1 INTRODUCTION

Having discussed in previous chapters the question of CO_2 accumulation and the various types of environmental feedback which pollution can induce, we turn to the interrelations between atmospheric waste disposal, capital accumulation and economic growth. This is done within a simple modelling framework, there is a further extension of results in a more specific context of Chapter 8. After specifying a neoclassical growth model which incorporates both waste flows and pollution abatement, we check for equilibrium and balanced expansion paths before and after the aggregate level of pollution is socially regulated. Some of these theoretical conclusions about static and dynamic equilibrium are interpreted from a political economic perspective. A seminal classical work in the context of environmental pollution and economic growth is d'Arge's (1971), and our models relate structurally to his, although the focus here is clearly in the framework of general equilibrium.

One major conclusion is that, although some combination of physical constraints given by global pollution might eventually retard economic growth, there is no reason to believe that pollution alone would necessarily do so, which naturally leads to the concept of sustainable growth (Pezzey, 1989). Another major conclusion is that the implementation of a strict programme of pollution management will probably encounter strong political opposition as long as the population is divided and there is international disagreement.

Before proceeding to a complete specification of the dynamic model we will use to analyse pollution and growth, we first need to understand the variety of ways in which waste discharges can lead to real economic costs. These costs of pollution can be classified into two broad categories. First, many of the repercussions of environmental pollution share the property that they decrease the aggregate production possibilities of an economy at any moment of time, either by decreasing the supplies of appropriable factors of production or by decreasing their productivity. Second, environmental pollutants can also lead to real costs by their direct effects on the utility levels of people, holding constant the supplies of goods and services available to them.

Which of these categories of cost is important is, of course, an empirical question. In his influential The Costs of Economic Growth, Mishan (1967) makes a great deal of the latter category, e.g. the environmental disamenities associated with economic growth. Certainly, one would not want to ignore or discount the importance of these environmental disamenities associated with pollution. However, in the context of global pollution treated

215

in previous chapters, the effects of pollution on real production and trade might eventually become far more serious than these direct utility effects if pollution were to continue growing. It is for this reason that the present analysis focuses on the feedback effects of pollution on real production.

7.2 CUMULATIVE POLLUTANTS AND THE RATE OF ECONOMIC GROWTH

How do we study the effects of cumulative pollutants and pollution abatement on the rate of economic growth? In particular, we need to see under what conditions unrestrained waste emissions retard the growth rate of production. More important, we need to discover what impact strict regulation of the level of pollution would have on the growth rate of the economy.

Before we can assess the thesis that environmental pollution is an unavoidable limit to continued economic growth, we first need to construct an economic growth model which depicts the productive consequences of polluting the environment and the costs of treating and recycling wastes. The theoretical approach taken here stems from the earlier work of Solow (1956) and Uzawa (1963), in that it assumes aggregate production relations which permit substitution between homogeneous capital and labour inputs. Alternative conclusions may be obtained by departing from a neoclassical approach to growth and the environment (Hung et al., 1993).

Production takes place in two sectors, the first output being a homogeneous commodity and the second being the disposal of potentially harmful waste flows.

The commodity output can be used either as a durable investment good, or alternatively, as a non-durable consumer good.

When it is used as an investment good, the commodity output either replaces depreciated capital or augments the stock of depreciable capital.

The output of waste disposal, on the other hand, is measured by the physical quantity of homogeneous waste material purified, recycled or otherwise not discharged into the environment in potentially costly forms. Gross national product consists of current commodity production plus the current level of waste disposal activity valued in units of commodity output. This is equivalent to saying that GNP equals current consumptions, gross investment in physical capital, and intermediate production undertaken to maintain environmental quality.

Net national product equals current consumption and net capital accumulation less the current costs of pollution damages.

The level of pollution is assumed to affect production possibilities in several different ways. In the first place, the percentage rate of population growth and the number of manhours worked per capita both respond negatively to environmental pollution beyond some

217

threshold level of concentration. These two feedback effects can limit the supply of labour efficiency units available per period. Environmental pollution can also accelerate capital depreciation, thereby restricting the rate of capital accumulation.

In addition to these effects on factor supplies, the disposal of wastes in the natural environment can also reduce factor productivities by disrupting the flow of environmental services which complement these appropriable factors in production. As long as the level of pollution remains unchanged, production in both sectors displays constant returns to scale. However, after some critical threshold, a growing level of pollution results in decreasing returns to the scale of productive activity, a proposition which will be demonstrated below. For analytical convenience, it is assumed that rising pollution does not affect the ratio of marginal products within a sector for any particular factor endowment.

Having made these assumptions, one can begin to sketch a neoclassical model of economic growth and environment pollution. Turning first to relations between population, labour supply and employment, we specify that:

$$\dot{N}/N = n(P) , \quad dn/dP \leq 0, \tag{1}$$

where N is the total population of the society and P is the level of environmental pollution.

$$M = \phi(P) \cdot N, \quad 0 < \phi < 1, \quad d\phi/dP \leq 0, \tag{2}$$

where M is the number of manhours offered per period and ϕ is a factor which embodies the combined effects of P on the labour force participation rate and average manhours offered per worker per period.

$$L = M \cdot e^{mt}, \quad m > 0, \tag{3}$$

where L is the number of labour-efficiency units available for production and m is the percentage rate of Harrod-neutral technical progress, i.e,. technical innovation augments the productivity of each manhour worked by a rate of m percent per period.

218

$$L = L_1 + L_2, \tag{4}$$

where L is the total effective labour supply and L_1 and L_2 are the labour forces employed in commodity production and pollution abatement, respectively.

The technical relations of production in the two sectors are specified by the following aggregate production functions:

$$Q_1 = \beta_1(P) \cdot F_1(K_1, L_1), \quad d\beta_1/dP \leq 0, \tag{5}$$

where Q_1 is total commodity production and K_1, the capital employed in its production. β_1 is an indirect measure of the level of environmental services to commodity production. F_1 displays constant returns to scale.

$$Q_2 = \beta_2(P) \cdot F_2(K_2, L_2), \quad d\beta_2/dP \leq 0, \tag{6}$$

where Q_2 is the physical quantity of homogeneous wastes which are treated, recycled or otherwise not discharged in destructive forms, and where K_2 is the capital devoted to pollution abatement. F_2 is also linearly homogeneous.

Whether or not production in both sectors actually displays constant returns is, of course, an empirical question. The assumption of linear homogeneity is adopted here for the sake of analytical convenience but especially because case studies of a number of particular industries have revealed constant returns in production.

The theoretical linkages among saving, investment and capital accumulation take the following forms:

$$Q_1 = C + I, \tag{7}$$

where C is current consumption of, and I is gross investment from, commodity output.

$$I = \dot{K} + D, \tag{8}$$

219

where K is net capital accumulation and D is depreciation of the existing capital stock.

$$D = \delta(P) \cdot K, \quad d\delta/dP \geq 0, \quad 0 < \delta < 1, \tag{9}$$

with δ the proportionate rate of capital depreciation and K the available stock of capital.

Full employment of capital requires that

$$K = K_1 + K_2, \tag{10}$$

and the identity of gross investment, I, and gross saving, S, dictates that

$$I = S, \tag{11}$$

The savings function is a simple proportional relation between saving and gross income, Y:

$$S = s \cdot Y, \quad 0 < s < 1. \tag{12}$$

The determination of prices, wages, profits and income requires that we add still another set of relationships to our model of pollution and growth. From the perspective of gross production,

$$Y = Q_1 + p \cdot Q_2, \tag{13}$$

where p is the price of abatement relative to the price of commodities and where the commodity price is the numéraire.

By this definition, gross product is measured in units of commodity production. From the point of view of income generation,

$$Y = v \cdot L + r \cdot K + D, \tag{14}$$

220

where v is the wage rate in commodity units and r is the rate of profit on capital.

The ratio of factor prices is determined exogenously, a specification which will be discussed below at some length:

$$z = v/(r + \delta).$$ (15)

The relative product price, p, on the other hand, is defined as the ratio of unit costs of production of the two outputs. If we assume that the wage and profit rates are the same in both sectors, then the relative product price is determined by the factor price ratio and the allocation of factors between the two productive sectors:

$$p = \frac{[v \cdot L_2 + (r + \delta) \cdot K_2]/Q_2}{[v \cdot L_1 + (r + \delta) K_1]/Q_1}$$ (16)

$$= \frac{z \cdot L_2 + K_2}{z \cdot L_1 + K_1} \cdot \frac{Q_1}{Q_2}$$

The last set of relations in the model describes how waste flows are generated and poses three alternative specifications of the relationship between current waste flows, current abatement activity and the level of environmental pollution:

$$R = h_1 \cdot Q_1 + h_2 \cdot Q_2 + h_3 \cdot C,$$ (17)

where R is the current flow of waste residuals generated by economic activity and $h_1 > 0$, $h_3 < 0$, $h_2 < 1$. The constant $(1 - h_2)$ is the net amount of wastes abated per unit of abatement activity after one takes into account the waste residuals which pollution abatement itself creates.

The alternative definitions of environmental pollution are:

$$P = R - Q_2 - a,$$ (18)

where P is specified as that portion of R not currently treated, recycled or naturally assimilated. The constant flow of natural assimilative capacity is equal to a. By this definition, wastes do not persist in their harmful forms but, rather, decay fully after their

221

discharge.

$$\dot{P} = R - Q_2 - \alpha \cdot P, \tag{19}$$

where P is a persistent stock of wastes which accumulates in the environment and which decays at some proportionate rate, α.

$$\dot{P} = R - Q_2 - A(P), \quad dA/dP \leq 0, \tag{20}$$

where P is a stock of wastes subject to a limited flow of biodegradation per period, a process which itself deteriorates as pollution rises above some threshold level.

7.3 ECONOMIC GROWTH IN AN UNREGULATED ECONOMY

During its initial phase of unfettered growth, our model has a structure considerably simpler than that in the general case. Since $Q_2 = K_2 = L_2 = 0$ and since parameter values do not yet vary with the value of P, the model takes the following form:

$$L = \bar{\phi} \, N_o \cdot e^{(m + \bar{n}) t} \tag{3'}$$

$$Y = \bar{\beta}_1 \cdot F_1 (K, L) \tag{5'}$$

$$Y = C + I \tag{7'}$$

$$I = \dot{K} + D, \tag{8}$$

$$D = \bar{\delta} \cdot K, \tag{9'}$$

$$I = S, \tag{11}$$

$$S = s \cdot Y, \tag{12}$$

$$Y = v . L + r . K + D, \tag{14}$$

$$R = h_1 \cdot Y + h_3 \cdot C, \tag{17'}$$

$$P = R - a, \tag{18'}$$

$$\dot{P} = R - \alpha \quad \text{and, finally,} \tag{19'}$$

$$\dot{P} = R - \bar{A}. \tag{20'}$$

In this simplified one-sector case, the analysis is nearly identical to that pursued by Solow in his influential 1956 essay on steady state growth. According to Stiglitz and Uzawa (1969, p.6):,

> Steady states are the generalization to economies with growing population of the equilibrium analysis of stationary states. In a steady state, output, population, capital stock, consumption, and investment are all growing exponentially, while the capital-output ratio, interest rate, and consumption-investment ratio are constant. Because the proportions in which the different commodities are produced are constant (the consumption-investment goods ratio), steady states are often referred to as <u>balanced growth paths</u>.

Demonstrating the possible existence of a balanced growth path is relatively easy in this initial case. By substituting (9′), (11) and (12) into (8), we find that

$$\dot{K} = s \cdot Y - \overline{\delta} \cdot K. \tag{21}$$

If we define $k \equiv K/L$, multiply k by L and then differentiate with respect to time, we   also find that

$$\dot{K} = \dot{k} \cdot L + k \cdot \dot{L}. \tag{22}$$

Differentiating (3) with respect to time and substituting into (22), we know that

$$\dot{K} = \dot{k} \cdot L + k \cdot (m + \overline{n}) \cdot L. \tag{23}$$

Because of the earlier constant returns to scale assumption about F_1, we can use $Y \equiv Y/L = y(k)$ to set (21) equal to (23) and then divide by L:

$$\dot{k} = s \cdot y(k) - (m + \overline{n} + \overline{\delta}) \cdot k. \tag{24}$$

This relationship is particularly useful because it allows us to check whether there is some particular capital-labour ratio, call it k*, for which dK/dt = 0. If so, then full-employment balanced growth can take place: a fixed k would mean that capital and labour, and hence gross output, were all growing at the same proportional rate. It follows immediately that output per labour unit and output per unit of capital would also be fixed in value.

We can see that there is indeed the possibility of an equilibrium k value, k*, by graphing (24) as the vertical difference between its two terms, as in Figure 1. Because of the fairly strong diminishing returns to increasing the capital-labour ratio, the value of y grows less and less rapidly as k increases. This ensures that the two curves intersect and that a full-employment equilibrium value of k exists. This equilibrium is also stable, since $\dot{k} > 0$ when k < k*, and vice versa.

Along this balanced growth path, income must grow at a percentage rate which is just high enough to satisfy investors that the current output-capital ratio justifies accumulating capital at the same percentage rate. This warranted rate of economic growth, denoted G_W, is given by

$$G_w = s \cdot \frac{Y}{K} - \overline{\delta} = s \cdot \Omega - \overline{\delta}, \tag{25}$$
$$\text{where} \quad \Omega \; \bullet \; Y/K = y(k)$$

This requirement is not sufficient, however, to ensure balanced growth of the economy. In addition, the maintenance of full employment of labour requires that income also grow at the natural rate of increase of labour supply, denoted G_n.

$$G_n = m + \overline{n}. \tag{26}$$

Since both of these requirements must be satisfied at the same time, the complete requirement for balanced growth with full employment is that,

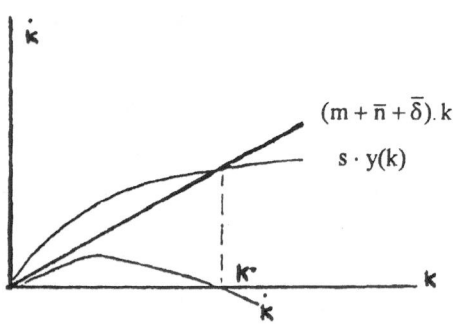

Figure 1

Existence of Balanced Growth
in an Unregulated Economy

$$\frac{\dot{Y}}{Y} = s \cdot \Omega^* - \overline{\delta} = m + \overline{n},$$ (27)

where $\Omega^* = y(K^*) / K^*$.

Since $\dot{Y}/Y = m + \overline{n}$ and since consumption is proportional to gross income, total consumption also grows at a percentage rate of $(m + \overline{n})$. This means that per capita consumption grows at an exponential rate of m percent per period, as do per capita income and output per manhour.

Up to this point, all of our results comply exactly with Solow's earlier findings. Economic growth entails more than rising incomes and consumption, however, especially if one considers the question of waste disposal. By substituting (7') and (12) into (17'), we find that the current volume of waste residuals generated by production and consumption is proportional to gross income:

$$R = [h_1 + (1 - s) \cdot h_3] \cdot Y.$$ (28)

226

Thus, along the equilibrium growth path, the flow of current waste discharges grows at the same exponential rate as gross income, namely (m + n̄) percent per period.

Since no pollution abatement activities are undertaken during this initial stage of economic growth, pollution also grows unchecked, except for those wastes assimilated by natural processes. Whether one regards wastes as persistent or not, this means that the level of pollution grows even faster than gross income. We can see this result by considering the three alternative definitions of environmental pollution.

When wastes are not persistent and are assimilable only up to a point by natural processes, the following relationship holds along the equilibrium growth path once $P = R - a > 0$:

$$\frac{\dot{P}}{P} = \frac{\dot{R}}{R - a} = \frac{(m + \bar{n}) \cdot R}{R - a} = \frac{(m + \bar{n})}{1 - (a/R)} . \tag{29}$$

Since $1 > (a/R) > 0$, it is clear that $\dot{P}/P > (m + \bar{n})$. Because the flow of waste residuals grows compared to the natural assimilative capacity of the environment, a/R falls as time passes. Thus, $\dot{P}/P$ declines in value toward (m + n̄) although it certainly does not approach (m + n̄) as a limit since the stage of growth ends long before a/R approaches zero. (See Figure 2).

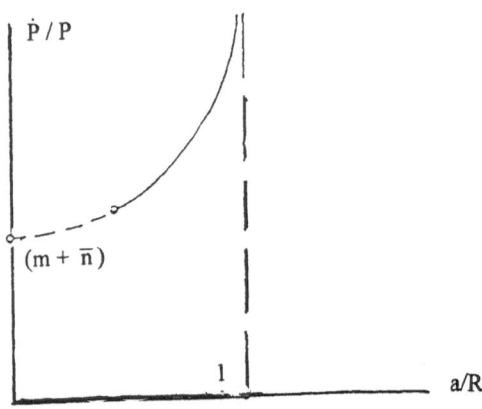

Figure 2

Rate of Growth of Pollution
During Stage of Development in Unregulated Economy
When $P = R - a > 0$

This relatively rapid growth of pollution has an important and immediate implication, namely that P/Y increases as the economy grows. This result is tentatively confirmed by results, reported by Daly and Cobb (1989), that for most industrialized countries change in pollution level has significantly and consistently exceeded GNP growth for this long period.

A reasonable explanation for this apparently rising "propensity to pollute" is that a variety of particular technologies adopted during postwar periods have replaced ecologically less disruptive techniques of production. Synthetic detergents, for example, have largely supplanted fat-based soaps. To a substantial extent, synthetic fibres have substituted for cotton and wool in the production of textiles. Pesticides, herbicides and nitrogen fertilizers have largely replaced manure and natural pest controls in worldwide agriculture.

The significance of equation (29) is that it permits one to explain the rising ratio of pollution to GNP without appeals to sectoral shifts in production or to changes in production techniques within industries. The mere fact that the waste generated in production and consumption has outstripped the natural capacity of the environment to assimilate wastes would account for that outcome. Actually, some combination of these two factors is probably responsible for the relatively rapid growth of pollution during recent decades.

Qualitatively identical results hold when wastes are persistent but decay at some constant proportional rate. That is, when $P = R - \alpha P$, the level of environmental pollution grows faster than $(m + \bar{n})$ percent per period. In order to reach this result, multiply (19) by $e^{\alpha t}$:

$$\dot{P} \cdot e^{\alpha t} + \alpha \cdot P \cdot e^{\alpha t} = R \cdot e^{\alpha t}. \tag{30}$$

Since the left-hand side of this relation is equal to $d(Pe^{\alpha t})/dt$, we can take the antiderivative of (30) to obtain

$$P(t) \cdot e^{\alpha t} = \int_0^t R(t) \cdot e^{\alpha t} \, dt. \tag{31}$$

Along the equilibrium growth path, $R(t) = R_o \cdot e^{(m+\bar{n})}$ so that

$$
\begin{aligned}
P(t) \cdot e^{\alpha t} &= R_o \cdot \int_0^t e^{(\alpha + m + \bar{n})t} \, dt \\
&= \frac{R_o}{\alpha + m + \bar{n}} \cdot [e^{\alpha + m + \bar{n})t} - 1].
\end{aligned} \tag{32}
$$

Dividing by $e^{\alpha t}$ and taking the time derivative of P, we note that

$$\dot{P} = \frac{R_o}{\alpha + m + \bar{n}} \cdot [(m + \bar{n}) \cdot e^{(m+\bar{n})t} + \alpha \cdot e^{-\alpha t}]. \tag{33}$$

Dividing once more, this time by P, we finally get the percentage growth rate of pollution along the equilibrium path:

$$\frac{\dot{P}}{P} = \frac{(m + \bar{n}) + \alpha \cdot e^{-(\alpha + m + \bar{n})t}}{1 - e^{(-\alpha + m + \bar{n})t}} \tag{34}$$

Clearly, when $\dot{P} = R - \alpha \cdot P$ pertains, pollution grows faster than $(m + \bar{n})$ percent annually along the equilibrium path, but this growth rate falls toward $(m + \bar{n})$ as balanced

229

growth proceeds.

Not surprisingly, the third definition of environmental pollution leads to these same conclusions about the growth of pollution during the stage of development. When $\dot{P} = R - \bar{A}$, it is generally true that

$$P(t) = \int_o^t (R - \bar{A}) \, dt. \tag{35}$$

Along the balanced growth path,

$$P(t) = R_o \cdot \int_o^t e^{(m + \bar{n}) t} - \bar{A} \cdot t. \tag{36}$$

Defining $R_o \equiv \bar{A}$ so that we can focus on the period after wastes begin to accumulate, we find that

$$P(t) = \bar{A} \cdot \left[\int_o^t e^{(m + n) t} \, dt - t \right]$$
$$= \bar{A} \cdot \left[\frac{1}{m + \bar{n}} \cdot [e^{(m + \bar{n}) t} - 1] - t \right]. \tag{37}$$

Calculating the percentage growth rate of P along this equilibrium path, we discover that

$$\frac{\dot{P}}{P} = \frac{1}{\dfrac{1}{m + \bar{n}} - \dfrac{t}{e^{(m + \bar{n}) t - 1}}} \tag{38}$$

As in the previous two cases, $\dot{P}/P$ exceeds $(m + \bar{n})$ and also falls toward that value as balanced growth takes place. Our conclusion is that, regardless of the particular definition of pollution one adopts, pollution grows faster than gross production during stages of balanced growth in an unregulated economy.

7.4 ECONOMIC GROWTH IN A DEGRADING ECONOMY

At some point in history, the uninhibited discharge of economic wastes into the natural environment finally begins to tax the assimilative capacity of the environment and result in costly environmental repercussions. We shall call the period when pollution abatement measures have not yet been adopted, "the degrading economy."

Once this degrading period of economic growth has been reached, it becomes substantially more difficult to analyse the dynamic behaviour of the economy. The reason is that a number of erstwhile parameters affecting labour force participation, the rate of capital depreciation, the rate of population increase, and the flow of environmental services begin to vary with the level of pollution. Consequently, it becomes much more complicated to predict the rate of growth of labour supply and the rate of capital accumulation and, hence, the rate of income growth.

During this transitional period, our general model of the environment and aggregate economy reduces to the following system of relationships:

$$\dot{N}/N = n(P) , \quad dn/dP < 0, \tag{1'}$$

$$M = \phi(P) \cdot N, \quad 0 < \phi < 1, \quad d\phi/dP < 0, \tag{2'}$$

$$L = M \cdot e^{mt}, \quad m > 0, \tag{3'}$$

$$Y = \beta_1(P) \cdot F_1(K, L), \quad d\beta_1/dP < 0, \tag{5"}$$

$$Y = C + I, \tag{7'}$$

$$I = \dot{K} + D, \tag{8}$$

$$D = \delta(P) \cdot K, \quad d\delta/dP > 0, \quad 0 < \delta < 1, \tag{9"}$$

$$I = S \tag{11}$$

$$S = s \cdot Y, \quad 0 < s < 1, \tag{12}$$

231

$$Y = v \cdot L + r \cdot K + D, \tag{14}$$

$$z = v/(r + \delta), \tag{15}$$

$$R = h_1 \cdot Y + h_3 \cdot C, \tag{17'}$$

$$P = R - a, \tag{18'}$$

$$\dot{P} = R - \alpha \cdot P, \quad \text{and, finally,} \tag{19'}$$

$$\dot{P} = R - A(P), \quad dA/dP{<}0. \tag{20''}$$

One way to measure the economic repercussions of pollution during this period is to compare the level of commodity production which present factor supplies can produce with the aid of current environmental services to that level which they could have produced if environmental services had not deteriorated. Let us define $E(P) = \beta_1(P)/$ where E is the measure of environmental services. We can then write the production relation (5") as

$$Y = E(P) \cdot \overline{\beta}_1 \cdot F_1(K, L) = E(P) \cdot \overline{Y}. \tag{39}$$

This formulation represents current production as some "primal" level of output adjusted for past decreases in environmental services.

One consequence of this sensitivity of environmental services to the level of pollution is that commodity production displays diseconomies of scale with respect to labour and capital during the degrading era. Expressing (39) in percentage growth terms, one finds that

$$\frac{\dot{Y}}{Y} = (\eta_{F_1 \cdot K}) \cdot \frac{\dot{K}}{K} + (\eta_{F_1 \cdot L}) \cdot \frac{\dot{L}}{L} + (\eta_{E \cdot P}) \cdot \frac{\dot{P}}{P}, \tag{40}$$

where η_{ij} is the elasticity of i with respect to j.

In the absence of the term involving environmental services and pollution, equality of the rates of growth of labour and capital would imply that gross production grew at that same common rate since F_1 is linear and homogeneous. However, since $\eta_{E\cdot P} < 0$ and $\dot{P}/P > 0$, gross income has to grow less rapidly than do the appropriated factors of production. This diseconomy of scale reflects the continuing deterioration of environmental quality which results from growing concentrations of environmental pollution.

The consequences of unrestrained waste discharges extend beyond these immediate effects on factor productivity, however. They also include substantial impacts on the rate of capital accumulation. If we substitute (5"), (9"), (11), and (12) into (8) and then divide by K, we find

$$\frac{\dot{K}}{K} = s \cdot E(P) \cdot \overline{\Omega}(k) - \delta(P), \tag{41}$$

where $\overline{\Omega} = (\overline{Y}/K) = \overline{y}(k)/k$

What this result points out is that, for any particular value of k, the higher the level of pollution, the lower the percentage rate of capital accumulation. This retardation of capital accumulation reflects two separate environmental repercussions of pollution. First, deteriorating environmental services reduce the ability to produce commodities for either capital replacement or accumulation purposes. Second, the proportion of the capital stock which needs replacement each period increases, thereby decreasing the portion of gross investment available for accumulation.

The effect of environmental pollution on the growth of labour supply is even more interesting, analytically speaking, than that on the rate of capital accumulation. The proportional rate of growth of labour supply varies inversely, not just with the level of pollution, but also with the growth rate; and the number of labour units worked per capita vary with the level of pollution, we can see that

$$L = \phi(P) \cdot N_o \cdot e^{[m + n(P)] \cdot t}. \tag{42}$$

If one calculates the percentage rate of growth of L, it becomes apparent that

$$\frac{\dot{L}}{L} = m + n(P) + [\eta_{\phi \cdot P} + \eta_{n \cdot P} \cdot nt] \frac{\dot{P}}{P}, \tag{43}$$

where $\eta_{\phi \cdot P} < C$ and $\eta_{n \cdot P} < 0$.

This finding means that the rate of growth of labour supply will be reduced if the level of pollution is relatively high because it retards the rate of population increase. In addition, however, if the level of pollution is growing as well as being absolutely high, then labour growth will be further reduced by the consequent reductions in hours worked per capita and in the population growth rate. This possibility is depicted in Figure 3, which shows the separate effects of P and $\dot{P}/P$ on the rate of labour increase.

Unfortunately, predicting the rate of income growth during this transitional period is impossible without detailed knowledge about parameter shifts and about the disequilibrium behaviour of the system. For one thing, a balanced growth path does not exist during the degrading era: even if labour and capital happened momentarily to grow at the same rate, income would grow less rapidly because of the diseconomies brought about by deteriorating environmental services.

One can say, however, that the warranted rate of economic growth during the degrading era is lower than it would have been during the unregulating era for the same value of k. To confirm this point, we need only recognize that

$$G_w = \frac{\dot{K}}{K} = s \cdot E(P) \cdot \overline{\Omega}(k) - \delta(P), \quad \text{whereas} \tag{44}$$

$$\overline{G}_w = s \cdot \overline{\Omega}(k) - \overline{\delta}. \tag{45}$$

234

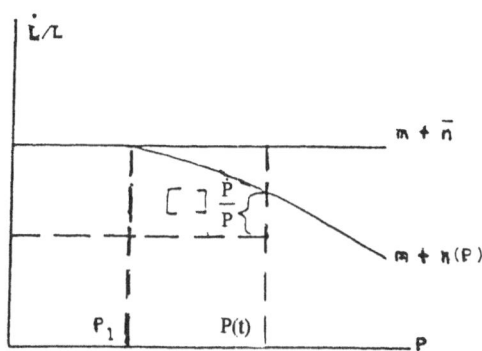

Figure 3
Relation between Rate of
Population Increase and Both the
Level and Growth of Pollution
During Degrading Era

In order to measure the cumulative, negative effect of rising pollution on the warranted rate of growth, we can take the difference between these two rates:

$$G_w - \overline{G}_w = s \cdot [E(P) - 1] \cdot \overline{\Omega}(k) + [\overline{\delta} - \delta(P)] < 0. \tag{46}$$

Although this point is useful, it still does not specify the actual disequilibrium behaviour of the economy during the degrading period. There is, however, a general dynamic tendency which we can identify. If the effects of rising pollution fall relatively heavily on the human factor, so that n and ϕ deteriorate sufficiently rapidly compared to E and δ, then $\dfrac{\dot{K}}{K} > \dfrac{\dot{L}}{L}$ As a result, the ratio of capital to labour will rise and, hence, $\overline{\Omega}$ will fall. This decline in $\overline{\Omega}$ acts as a brake on the rate of capital accumulation in addition to those reflecting rising capital depreciation and sparser environmental services.

If, on the other hand, the effects of rising pollution fall most severely on physical capital, then E and δ deteriorate relatively rapidly compared to n and ϕ, with the consequence that $\frac{\dot{K}}{K} < \frac{\dot{L}}{L} < m + \bar{n}$. In this event, k falls and $\bar{\Omega}$ rises. This increase in the "primal" productivity of capital acts to brake the retardation of capital accumulation, but the growth rate of the capital stock is nevertheless still below what it was during the unregulating era of balanced growth.

7.5 ECONOMIC GROWTH IN THE SPACESHIP ECONOMY

As we have just seen, the unregulating era of economic growth brings with it rising levels of pollution, deteriorating environmental services, reductions in factor supplies, and a lower warranted rate of growth. These effects of the unregulated discharge of effluents may not be prominent at first but, after some point, the ecological repercussions of waste emissions can no longer be ignored because of their increasing real cost. Because pollution grows faster than gross income and also because the various costs of pollution grow at least as fast as the pollution which causes them, waste discharges must eventually come under social regulation. Once this political decision to halt the growth of pollution has been reached, we have entered the era of the "spaceship economy" in the sense of Boulding (1968). Thereafter, sufficient quantities of capital and labour must be devoted to the new pollution abatement sector to prevent further losses of environmental quality.

Before turning to our analysis of the spaceship stage of economic growth, we must first re-specify our general model to take into account the fact that $Q_2 > 0$. The revised model looks as follows:

$$\dot{N}/N = n(P), \quad dn/dP<0, \tag{1'}$$

$$M = \phi(P) \cdot N, \quad 0<\phi<1, \quad d\phi/dP<0, \tag{2'}$$

$$L = M \cdot e^{mt}, \quad m>0, \tag{3}$$

$$L = L_1 + L_2, \tag{4}$$

$$Q_1 = \beta_1(P) \cdot F_1(K_1, L_1), \quad d\beta_1/dP<0, \tag{5'''}$$

$$Q_2 = \beta_2(P) \cdot F_2(K_2, L_2), \quad d\beta_2 \cdot /dP<0, \tag{6'}$$

$$Q_1 = C + I, \tag{7}$$

$$I = \dot{K} + D, \tag{8}$$

$$D = \delta(P) \cdot K, \quad d\delta/dP > 0, \quad 0 < \delta < 1, \tag{9"}$$

$$K = K_1 + K_2, \tag{10}$$

$$I = S \tag{11}$$

$$S = s \cdot Y, \tag{12}$$

$$Y = Q_1 + p \cdot Q_2, \tag{13}$$

$$Y = v \cdot L + r \cdot K + D \tag{14}$$

$$z = v/(r + \delta), \tag{15}$$

$$p = \frac{z \cdot L_2 + K_2}{z \cdot L_1 + K_1} \cdot \frac{Q_1}{Q_2}, \tag{16}$$

$$R = h_1 \cdot Q_1 + h_2 \cdot Q_2 + h_3 \cdot C, \tag{17}$$

$$P = R - Q_2 - a, \tag{18}$$

$$\dot{P} = R - Q_2 - \alpha \cdot P, \tag{19}$$

$$\dot{P} = R - Q_2 - A(P), \quad dA/dP < 0, \quad \text{and finally} \tag{20}$$

$$P = \tilde{P} \tag{47}$$

where $\tilde{P}$ is the policy ceiling on the level of pollution.

Having reformulated the model, our next step is to analyse what effect the imposition of $\tilde{P}$ has on the static equilibrium of the spaceship economy. The first point in this analysis is that, regardless of the particular value of $\tilde{P}$, any arbitrary value of the factor price ratio, call it z^*, implies a unique pair of cost-minimizing capital-labour ratios, k_1^* and k_2^*, in the

238

two production sectors. It is the twin assumptions that pollution affects the production relations only via the separable $\beta_i(P)$ co-efficients and that the F_i factors of the production functions are linear and homogeneous which ensure that these cost-minimizing capital-labour ratios depend only upon relative factor price and not upon the scale of production in the two sectors.

The aggregate capital-labour ratio, on the other hand, does depend on the particular choice of $\tilde{P}$. If we define $\tilde{L}$ as that labour force offered for employment when $P = \tilde{P}$, then

$$\tilde{L} = \phi(\tilde{P}) \cdot N \cdot e^{mt}, \quad d\tilde{L}/d\tilde{P} < 0. \tag{48}$$

By contrast, the current size of the capital stock does not vary with $\tilde{P}$ Drawing on the full employment assumptions (4) and (10), we also know that

$$\tilde{K} = \frac{K_1 + K_2}{L_1 + L_2} = \frac{K_1 + K_2}{\tilde{L}}$$
$$= \lambda \cdot k_1 + (1 - \lambda) \cdot k_2, \tag{49}$$

where $\tilde{K} = K/\tilde{L}$, $k_i = K_i/$ and $\lambda = L_1/\tilde{L}$.

That is, besides depending on $\tilde{P}$, the aggregate capital-labour ratio is the weighted average of the two sectoral ratios, where the weights are the fraction of the total labour force allocated to each sector.

Let us continue by tentatively assigning some particular value to $\tilde{P}$ and then seeing whether or not the pieces of our puzzle fall nicely into place. Given the aggregate $\tilde{k}$ corresponding to $\tilde{P}$ and given the two previous assumptions about full employment and cost minimization, then the unique k_i^* values corresponding to any arbitrary z^* determine a unique proportional allocation of labour between the two sectors:

$$\lambda^* = \frac{\tilde{K} - k_2^*}{k_1^* - k_2^*} \quad \text{and} \quad (1 - \lambda^*) = \frac{k_1^* - \tilde{K}}{k_1^* - k_2^*}. \tag{50}$$

Since the current size of the labour force in efficiency units is fixed by this pollution

239

ceiling, these proportions in turn uniquely determine the absolute endowments of labour units in both sectors:

$$L_1^* = \lambda^* \cdot \tilde{L}, \text{ and } L_2^* = (1 - \lambda^*) \cdot \tilde{L}.$$

(51)

The absolute endowments of capital, K_1^* and K_2^* are then implied by the unique cost-minimizing capital-labour ratios:

$$K_i^* = k_i^* \cdot L_i^*, \quad i=1,2.$$

(52)

Finally, since we have deduced the factor endowments in both sectors, we also know the specific production rates, Q_1^* and Q_2^* which correspond to $\bar{P}$ and the arbitrarily chosen z^*:

$$Q_i^* = \beta_i(\bar{P}) \cdot F_i(K_i^*, L_i^*), \quad i=1, 2$$

(53)

Once these sectoral factor endowments and production rates have been determined, it is also possible to calculate a unique relative price, p^* , for the two outputs. Recalling that p equals the ratio of their unit costs of production, we note that

$$p^* = \frac{z^* \cdot L_2^* + K_2^*}{z^* L_1^* + K_1^*} \cdot \frac{Q_1^*}{Q_2^*}$$

(54)

Having calculated this price, one can then easily find the values of gross income and saving which correspond to $\bar{P}$ and z^* . In particular,

$$Y^* = Q_1^* + p^* \cdot Q_2^*$$

(55)

and $\quad S^* = sY^*$

(56)

Calculating current consumption requires a bit more effort. By combining (7) and (11) - (13), we can express consumption as a function of relative product price and the two

240

production rates:

$$C^* = (1 - s) \cdot Q_1^* - s \cdot p^* \cdot Q_2^* \tag{57}$$

Somewhat surprisingly, this means that the average propensity to consume commodity output is not a constant but, rather, varies inversely with the ratio of the value of abatement to commodity production.

$$C^*/Q_1^* = (1 - s) - s \cdot \frac{p^* \cdot Q_2^*}{Q_1^*} < (1 - s). \tag{58}$$

Are we to conclude from all of this that a static equilibrium exists for the spaceship economy? Not necessarily. The reason is that the particular production rates and consumption flow implied by the existing pollution ceiling and factor price ratio, taken together with the waste propensities h_i, might tend toward a level of pollution greater than $\tilde{P}$. In that event, z^* and $\tilde{P}$ are incompatible values and a momentary equilibrium does not exist for them.

We can see this clearly in the case where (18) pertains by expressing the potential level of pollution as a function of Q_1^*, Q_2^*, and p^*. Substituting (17) and (57) into (18), we discover that

$$a + P^* = c_1 \cdot Q_1^* + c_2 (p^*) \cdot Q_2^*, \tag{59}$$

where $\begin{aligned} c_1 &= [h_1 + (1 - s) \cdot h_3] > 0, \\ c_2 (p^*) &= [(h_2 - 1) - s \cdot h_3 \cdot p^*] \end{aligned}$ and

P^* is that potential level of pollution which the existing factor prices, waste propensities, full employment, cost minimization and the factor supplies and environmental services corresponding to $\tilde{P}$ tend to lead to.

In general, there is no reason to expect that $\tilde{P}$ and z^* lead to a momentary equilibrium, i.e. that $P^* = \tilde{P}$. Rather, if the pollution ceiling and existing factor price ratio imply relatively little abatement activity and also a comparatively low product price ratio, so that a relatively large portion of commodity output is consumed, then P^* is likely to exceed $\tilde{P}$ If, on the other hand, z^* and $\tilde{P}$ suggest relatively little commodity production and a fairly low propensity to consume commodity output, then P^* is probably smaller than the ceiling initially hypothesized.

The same conclusion holds for those cases where wastes are persistent. In this event, the potential level of pollution corresponding to the existing factor price ratio and the factor supplies and environmental services implied by $P = \hat{P}$ and $\dot{P} = 0$ is either

$$\alpha \cdot P^* = c_1 \cdot Q_1^* + c_2 (p^*) \cdot Q_2^*, \quad \text{or} \tag{60}$$

$$A(P^*) = c_1 \cdot Q_1^* + c_2 (p^*) \cdot Q_2^*. \tag{61}$$

Again, there is no reason to expect that the production rates and price ratio which would correspond to some arbitrary z^* and our specified $\tilde{P}$ would actually result in a level of pollution equal to $\tilde{P}$ Hence, regardless of the definition of P , we should expect that arbitrarily chosen z^* and $\tilde{P}$ values will not result in momentary equilibrium of the economy. (The logic of this argument is summarized in Figure 4.)

In the event that a full equilibrium does not exist for the prescribed pollution ceiling and the current ratio of factor prices, one must then ask what adjustments, if any, would permit a full-employment equilibrium to emerge. An obvious candidate is changing the pollution ceiling so that the relative price and production rates implied by z^* and $\tilde{P}$ would, in turn, permit the ceiling to be actually realized. At first glance, one might expect that raising the policy ceiling would do the job. After all, if $P^* > \tilde{P}$ will not an increasingly high ceiling eventually equal P^* ? A little inspection reveals, however, that this is not generally the case. To see this point clearly, let us trace through the economic and environmental consequences of changing the pollution ceiling. First, drawing on (48), we note that

$$\frac{d\tilde{L}}{\tilde{L}} = \eta_{\phi \cdot \tilde{P}} \cdot \frac{d\tilde{P}}{\tilde{P}}, \tag{62}$$

where $\eta_{\phi \cdot \tilde{P}} = (d\phi/dP) \cdot (P/\phi) < 0$.

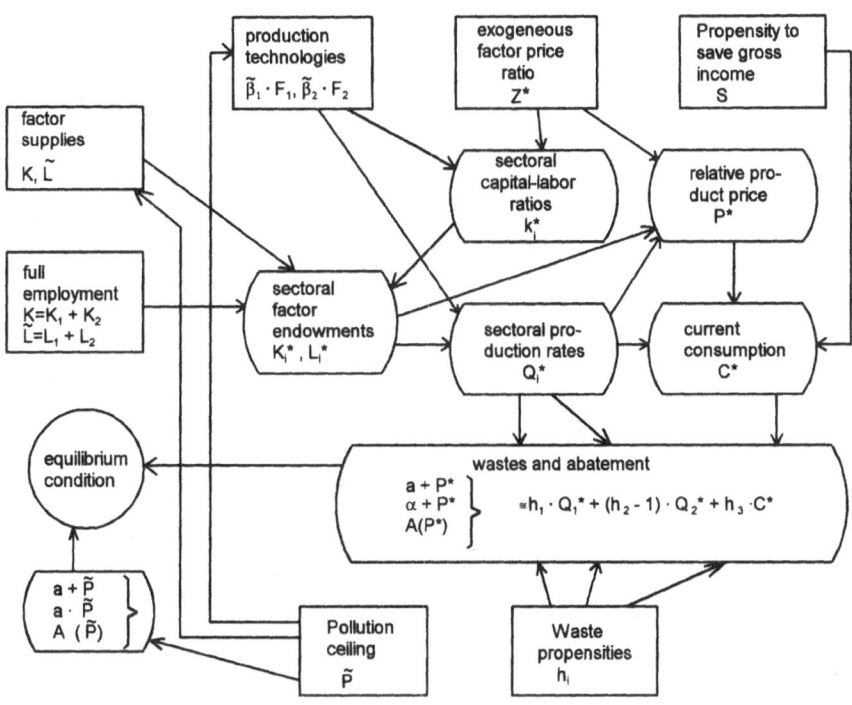

Figure 4

Production Rates, Consumption Flow
and Abatement Activity in Momentary Equilibrium

Since the capital stock is not immediately affected by the choice of $\tilde{P}$,

$$d\tilde{K} = -\tilde{K} \cdot \frac{d\tilde{L}}{\tilde{L}}, \quad \text{that is,} \tag{63}$$

$$d\tilde{K} = \frac{d\tilde{L}}{\tilde{L}}. \tag{64}$$

The reallocation of labour between commodity production and pollution abatement, on the other hand, depends upon both the direction of change in $\tilde{P}$ and also the relative capital intensities of the two sectors. From (50), we find that

$$d\lambda^* = \frac{1}{k_1^* - k_2^*} \cdot d\tilde{K} = \frac{\tilde{K}}{k_2^* - k_1^*} \cdot \frac{d\tilde{L}}{\tilde{L}} \tag{65}$$

and hence,

$$\frac{d\lambda^*}{\lambda^*} = \frac{\tilde{K}}{k_2^* - \tilde{K}} \cdot \frac{d\tilde{L}}{\tilde{L}}. \tag{66}$$

Similarly,

$$d(1 - \lambda^*) = -d\lambda^* \quad \text{and} \tag{67}$$

$$\frac{d(1 - \lambda^*)}{(1 - \lambda^*)} = \frac{\tilde{K}}{k_1^* - \tilde{K}} \cdot \frac{d\tilde{L}}{\tilde{L}}. \tag{68}$$

In other words, given the sectoral capital-labour ratios determined by the relative factor price, the smaller labour supply and consequently higher degree of aggregate capital intensity corresponding to a higher $\tilde{P}$ require that a larger portion of the labour force be allocated to the more capital intensive sector in order to maintain full employment. On the other hand, a more stringent $\tilde{P}$ dictates that a larger percentage of labour units be allocated to the more labour intensive sector.

Having demonstrated what effects changes in $\tilde{P}$ have on the availability of labour and its allocation between production sectors and also on the aggregate capital-labour ratio, we can now approach the question of what kinds of changes in $\tilde{P}$ are required to permit static equilibrium. In deciding this question, it is crucial to find out what impact a different $\tilde{P}$ would have on p^*, L and Q_i^*

We first observe that

$$Q_i^* = \beta_i(\tilde{P}) \cdot f_i(k_i^*) \cdot L_i^*, \quad i=1,2 \tag{69}$$

where $f_i(k_i^*) = F_i(K_i, 1)$

Since the k_i^* are fixed by z^* it follows that

$$\frac{dQ_i^*}{Q_i^*} = \frac{d\beta_i}{\beta_i} + \frac{dL_i^*}{L_i^*}, \quad i=1, 2. \tag{70}$$

The effects which different ceilings would have on total factor productivity in the two sectors are measured by

$$\frac{d\beta_i}{\beta_i} = \eta_{\beta_i} \cdot P \cdot \frac{d\tilde{P}}{\tilde{P}}, \quad i=1,2 \tag{71}$$

where $\eta_{\beta_i} \tilde{P} = \frac{d\beta_i}{d\tilde{P}} \cdot \frac{\tilde{P}}{\beta_i} < 0.$

The effects of different $\tilde{P}$ values on the absolute labour endowments in the production sectors are given by

$$\frac{dL_2^*}{L_2^*} = \frac{d\tilde{L}}{\tilde{L}} + \frac{d(1 - \lambda^*)}{(1 - \lambda^*)} \tag{72}$$

$$= (1 + \frac{\tilde{k}}{k_1^* - \tilde{k}}) \cdot \frac{d\tilde{L}}{\tilde{L}}$$

$$= (\frac{k_1^*}{k_1^* - \tilde{k}}) \cdot \eta_{\phi \cdot \tilde{p}} \cdot \frac{d\tilde{P}}{\tilde{P}} \quad \text{and}$$

245

$$\frac{dL_1^*}{L_1^*} = \frac{d\tilde{L}}{\tilde{L}} + \frac{d\lambda^*}{\lambda^*} \tag{73}$$

$$= (1 + \frac{\tilde{k}}{k_2^* - \tilde{k}}) \cdot \frac{d\tilde{L}}{\tilde{L}}$$

$$= (\frac{k_2^*}{k_2^* - \tilde{k}}) \cdot \eta_{\phi \cdot \tilde{P}} \cdot \frac{d\tilde{P}}{\tilde{P}}, \quad \text{respectively.}$$

From (70) - (73), it follows by substitution that

$$\frac{dQ_1^*}{Q_1^*} = \left[\eta_{\beta 1 \cdot \tilde{P}} + (\frac{k_2^*}{k_2^* - \tilde{k}}) \eta_{\phi \cdot \tilde{P}} \right] \cdot \frac{d\tilde{P}}{\tilde{P}} \quad \text{and} \tag{74}$$

$$\frac{dQ_2^*}{Q_2^*} = \left[\eta_{\beta 2 \cdot \tilde{P}} + (\frac{k_1^*}{k_1^* - \tilde{k}}) \eta_{\phi \cdot \tilde{P}} \right] \cdot \frac{d\tilde{P}}{\tilde{P}}. \tag{75}$$

What happens to the price ratio of outputs as $\tilde{P}$ varies is still another important question. Dividing the numerator and denominator of (54) by L_2 and L_1, respectively, we obtain

$$p^* = \frac{z^* + k_2^*}{z^* + k_1^*} \cdot \frac{q_1^*}{q_2^*}, \tag{76}$$

where $q_i = Q_i/L_i$, by definition.

From (69), we also know that

$$q_i^* = \beta_i(\tilde{P}) \cdot f_i(k_i^*), \quad i=1,2. \tag{77}$$

Hence, given the factor price ratio and the sectoral capital-labour ratios determined by it, the relative price of outputs varies with the ratio of environmental services to the two sectors:

$$p^* = \left[\frac{z^* + k_2^*}{z^* + k_1^*} \cdot \frac{f_1(k_1^*)}{f_2(k_2)} \right] \cdot \frac{\beta_1(\tilde{P})}{\beta_2(\tilde{P})}. \tag{78}$$

Whether p^* varies in the same direction as the pollution ceiling depends, then, on which production sector experiences more severe changes in environmental services. That is,

$$\frac{dp^*}{p^*} = [\eta_{\beta_1 \cdot \tilde{P}} - \eta_{\beta_2 \cdot \tilde{P}}] \cdot \frac{d\tilde{P}}{\tilde{P}} . \tag{79}$$

Having derived these incremental relations, we can finally ask what changes in $\tilde{P}$ might permit static equilibrium to occur. Expressed as an equilibrium condition derived from (59) - (61), this amounts to asking what value(s) of $\tilde{P}$ if any, would ensure that

$$G(\tilde{P}) = H(z^*, \tilde{P}) = c_1 \cdot Q_1^* + c_2(p^*) \cdot Q_2^* , \tag{80}$$

$$\text{where} \quad G(P) = \begin{cases} a + \tilde{P} \\ \alpha \cdot \tilde{P}, \quad \text{or} \\ A(\tilde{P}), \end{cases}$$

depending on which definition of P pertains.

From (74) and (75), it is clear that what happens to Q_1^* and Q_2^* and hence to H, as $\tilde{P}$ varies depends crucially on which sector is the more capital intensive. Let us assume for later analytical convenience that either $k_1 > k > k_2$ or $k_2 > k > k_1$ for all possible factor price ratios and then test these two cases for momentary equilibrium.

7.6 DYNAMIC EQUILIBRIUM IN THE SPACESHIP ECONOMY

Having looked at the static equilibrium effects of varying the factor price ratio, we can now inspect some of the dynamic properties of this model of environmental pollution and economic growth. In particular, one wants to know whether balanced growth can occur once some static equilibrium values of the pollution ceiling and the factor price ratio have been realized and, if so, whether that balanced growth path is stable.

A first starting point of this analysis is the observation that

$$\dot{K} = I - \delta \cdot K. \tag{81}$$

Substituting (11) - (13) into this fundamental dynamic relation, one obtains

$$\dot{K} = s \cdot Q_1 + sp \cdot Q_2 - \delta \cdot K. \tag{82}$$

Let us now define $c_o \equiv \begin{cases} a + \tilde{P} \\ \alpha \cdot \tilde{P}, \\ A(\tilde{P}), \end{cases}$

depending upon which definition of pollution we adopt. If we solve (59) - (61) for p, we get

$$p = b_1 \cdot (Q_1/Q_2) + b_2 \cdot (1/Q_2) + b_3, \tag{83}$$

where $b_1 = (h_1/sh_3) + (1-s)/s > 0,$

$$b_2 = - c_o/sh_3 < 0, \quad \text{and}$$

$$b_3 = (h_2 - 1)/sh_3 < 0.$$

Substituting this expression for p in (82), we find that

$$\dot{K} = s \cdot (1 + b_1) \cdot Q_1 + sb_3 \cdot Q_2 + sb_2 - \delta \cdot K, \tag{84}$$

Because of the linear homogeneity of the production functions, this can also be expressed as

248

$$\dot{K} = s \cdot (1 + b_1) \cdot q_1 \cdot \lambda \cdot L + sb_3 \cdot q_2 \cdot (1 - \lambda) \cdot L$$
$$+ sb_2 - \delta \cdot K$$
(85)

Using a favourite device of Solow, we know that
$$\dot{K} = \dot{k}L + k\dot{L}$$
$$= \dot{k}L + k \cdot (m + \tilde{n}) \cdot L$$
(86)

by differentiating K = kL with respect to time. Setting (85) equal to (86) and dividing by L, we find the time rate of change of k to be

$$\dot{k} = s \cdot (1 + b_1) \cdot q_1 \cdot \lambda + sb_3 \cdot (1 - \lambda) + \frac{sb_2}{L}$$
$$- (m + \tilde{n} + \delta) \cdot k.$$
(87)

Substituting (50) in (87), one obtains an expression for dk/dt which varies with the aggregate and sectoral capital-labour ratios:

$$\dot{k} = \left[-s(1+b_1) \cdot q_1 \cdot \frac{k_2}{k_1 - k_2} + sb_3 \cdot q_2 \cdot \frac{k_1}{k_1 - k_2} + \frac{sb_2}{L} \right]$$
$$+ \left[\frac{s \cdot (1 + b_1) \cdot q_1}{k_1 - k_2} - \frac{sb_3 q_2}{k_1 - k_2} \right] \cdot k - [m + \tilde{n} + \delta] \cdot k.$$
(88)

If we specify some factor price ratio, then the sectoral capital-labour ratios are fixed and so, therefore, are the first two bracketed expressions. The question then becomes whether there exists an aggregate k which could remain constant over time, i.e. a value of k such that $\dot{k} = 0$. If so, then there exists an economic growth path for the aggregate economy which is balanced and which satisfies the pollution constraint. As before, we need to inspect two alternative cases, that in which commodity production is relatively capital intensive and that in which it is relatively labour intensive.

When k_1 is greater than k_2, balanced growth may occur but its possibility is certainly not assured. This problematic result can be seen by graphing (88) as the difference between two linear equations. One equation consists of the first two bracketed terms and is positively sloped with a negative intercept. The second equation, which is subtracted vertically from the first to obtain $\dot{k}$ is a constant positive multiple of k.

These two linear equations intersect, thereby indicating the existence of some equilibrium value of the aggregate capital-labour ratio, k*, where $\dot{k} = 0$, only if the slope of the first equation exceeds $(m + \tilde{n} + \delta)$ To be specific, when $k_1 > k_2$, the existence of

249

balanced growth requires that

$$\frac{s \cdot (1 + b_1) \cdot q_1}{k_1 - k_2} - \frac{s \cdot b_3 \cdot q_2}{k_1 - k_2} > m + \tilde{n} + \delta.$$ (89)

Put more simply, the necessary condition for balanced growth equilibrium when commodity production is more capital intensive than abatement is that

$$\frac{(1 + \dfrac{h_1}{h_3}) \cdot q_1 - (\dfrac{h_2 - 1}{h_3}) \cdot q_2}{m + \tilde{n} + \delta} > k_1 - k_2.$$ (90)

One notable characteristic of this equilibrium requirement (90) is that the existence of balanced growth is independent of the savings ratio, a result common to several other neoclassical growth analyses. Another interesting result is that, even when it exists, equilibrium growth is unstable. As one can see in Figure 5, $\dot{k} > 0$ when $k > k^*$, and vice versa. Therefore, k tends to diverge from k^* unless k^* has already been attained.

What is especially surprising about (90), however, is the fact that a particular pair of $\tilde{P}$ and z^* values which do permit momentary equilibrium might not allow balanced expansion of the economy over time. That is, the specific values of the pollution ceiling and factor price ratio must be consistent, not only with full employment, of existing factor supplies, but also with capital accumulation sufficiently rapid to replace depreciated capital and to equip an expanding labour supply. If the price ratio and pollution ceiling are such that the situation in Figure 6 holds, then gross investment cannot occur rapidly enough at any capital-labour ratio to balance depreciation and growth of labour supply. Consequently, the aggregate capital-labour ratio tends to fall over time. Hence, when $k_1 > k_2$, the pairs of $\tilde{P}$ and z^* values which permit both momentary equilibrium and balanced expansion are fewer in number than those which permit static equilibrium alone.

In the case where abatement is always more capital intensive than commodity production, the situation is far simpler. Balanced growth is assured, regardless of the particular value of z, and this equilibrium path is stable.

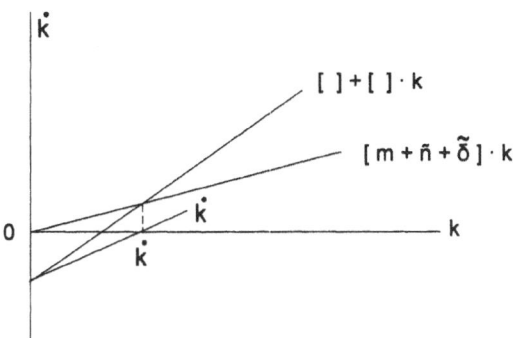

Figure 5

Existence of Dynamic Equilibrium
When Q_1 More Capital Intensive
and Necessary Condition Satisfied

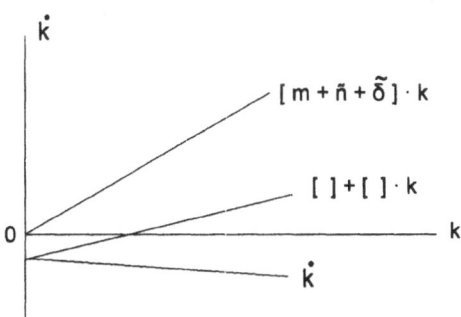

Figure 6
Non-Existence of Dynamic Equilibrium
When Q_1 More Capital Intensive
and Necessary Condition Unsatisfied

251

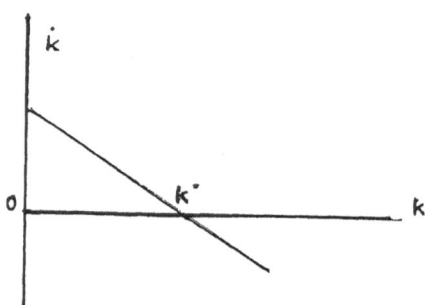

Figure 7
Existence of Dynamic Equilibrium
When Q_2 More Capital Intensive

This result is evident if we again graph (88) as in Figure 7. Since $(k_1 - k_2) < 0$, the first bracketed term is positive whereas the second bracketed term has a negative coefficient. Hence, (88) graphs as a negatively sloped linear equation with positive intercept for any particular value of z^*. This means that there is necessarily some equilibrium value of k, k^*, which also happens to be stable since $\dot{k} < c$ when $k > k^*$, and vice versa.

In conclusion, balanced expansion of the economy is compatible with maintaining a policy ceiling on environmental pollution if the pollution ceiling and the factor price ratio assume appropriate values. When pollution abatement is the more capital intensive activity, these values must simply ensure that some momentary equilibrium of the economy exists.

252

That is, the pollution ceiling and factor price ratio must permit full employment of both factors, equality of savings and investment, and the adoption of cost-minimizing techniques of production. When commodity production is more capital intensive, on the other hand, balanced expansion requires in addition that the pollution ceiling and factor price ratio allow a rate of capital accumulation rapid enough to match the growth of labour supply.

7.7 AN EVALUATION OF THE MODEL AND SOME CONCLUSIONS

Several characteristics of this model are noteworthy. It embodies a number of fairly optimistic assumptions about the economic impact of environmental pollution. The assumption, for example, that the current rate of population increase depends strictly on the current intensity of pollution is clearly a major simplification.

More realistically, the current rate of population growth reflects both the current and the previous levels of human exposure to harmful waste materials. Hence, if pollution levels have recently been rising, then the currently observed rate of population increase is greater than those rates which would prevail in the future even if pollution stabilized at the current level.

Precisely the same argument applies to the average number of manhours supplied per capita. Current morbidity and disability rates reflect both current and prevailing levels of exposure to pollution. Thus, if pollution has been growing in the recent past, then the currently observed value of ϕ exceeds those values which one would observe in the future even if pollution were to stop growing.

This model is also quite optimistic in its assumption that the rate of technical progress will be unaffected by measures taken to protect environmental quality. Whereas labour saving innovations have been adopted in the past almost without regard to their environmental side effects, it becomes necessary to reject certain technological possibilities once social control of waste discharges has become an economic and ecological imperative. Hence, it is dangerous to casually project past rates of technical progress into the future when they were obtained, at least in part, by adopting particular techniques which intensified the side effects of economic activity.

It does not follow from historical examples, of course, that rapid technological progress and careful attention to the ecological repercussions of new technologies are incompatible. It may be that the opportunities for fruitful research and development are still enormous even after one discards ecologically disruptive technologies. At a minimum, however, we can conclude that continued technical progress will require that future research efforts be more selective. The criteria for developing and adopting new techniques must be broadened to include their effects on waste propensities, toxicity and persistence as well as the effects they would have on factor productivity if there were no losses of environmental services associated with their introduction.

Still another sense in which this model is optimistic about continued growth is that it does not address the question of natural resource availability in a comprehensive manner. The effects of pollution on the availability of environmental services and biologically renewable resources are included, at least implicitly, in the two production functions of the model. The future availability of non-renewable ores and minerals, however, is omitted from consideration. As a result of this omission, any conclusions about the feasibility of continued economic growth should be interpreted to mean that environmental pollution alone is unlikely to halt the process of growth. This proviso leaves open the question whether spatial congestion, energy scarcity, the limited availability of arable land, and environmental pollution, taken as a set of physical constraints, might eventually prevent economic growth.

A final sense in which this analysis might be unduly optimistic is in its assumption that the production of pollution abatement enjoys constant returns to scale. Although this theoretical premise is a plausible one, its veracity awaits empirical confirmation as abatement activity expands in the coming years. If it turns out, however, that waste treatment and recycling actually display decreasing returns to scale then our conclusion that balanced growth and pollution control are compatible might not hold. In that event, the rising incremental expense of disposing of additional wastes would eventually retard economic growth which is in the mainstream of economic thinking (Daly, 1987).

Not all of the empirical assumptions incorporated in this model favour the conclusion that economic growth and the maintenance of environmental quality are compatible goals. One sense in which the analysis is unduly pessimistic is its assumption that flows of production or consumption, on the one hand, and the consequent flow of waste materials and energy, on the other hand, are fixed in proportion.

255

7.8 SOME LONG-TERM IMPLICATIONS OF THE MODEL

Beyond these predictions about the immediate effects of restraining environmental pollution, one can also draw some conclusions about long-term economic growth and the control of pollution. Perhaps the most basic of these results is that there is no physical contradiction in the long run between freezing the level of environmental pollution and maintaining a positive rate of economic growth. Balanced growth may require that relative factor prices change sufficiently to induce the selection of production techniques compatible with dynamic equilibrium. Once these techniques of production haven been adopted, however, both capital and labour can grow indefinitely at the proportionate rate, $m + \bar{n}$.

The fact that economic growth can proceed indefinitely despite the imposition of a ceiling on pollution means that the marginal cost of pollution abatement does not rise to prohibitive levels in the long run. We find that the ratio of pollution to both gross income and commodity production can approach zero along the equilibrium path. Thus, if one thinks of pollution control as a physical problem of accumulating enough additional capital in each time period to handle the extra waste loads resulting from economic growth, then the task of maintaining environmental quality amidst economic expansion can clearly be solved.

Whether or not the growth of pollution will actually be curbed is not, however, simply a matter of physical feasibility. It will also require that appropriate techniques of production be adopted and that a sufficiently large portion of productive resources be devoted to waste treatment and recycling. If productive capital were socially owned and the rental rate on capital were merely a shadow price for accounting purposes, then the changes in relative factor prices necessary to realize this programme of environmental protection would be of little political consequence. However, since the wage-rental ratio is a determinant of functional income shares in most contemporary economies, the implementation of a strict programme of pollution management will probably engender strong political opposition from either workers or owners of capital.

If enforcement of a pollution ceiling called for a substantially lower ratio of the wage to the rental on capital, so that the relative share of earnings fell in the short run, many workers might oppose implementation of that ceiling. This opposition would be particularly intense if there appeared to be no alternative to redistributing gross income from workers to owners other than abandoning the earlier commitment to full employment. If, on the other hand, the gross return to capital did not accrue to a largely distinct social class of owners but

256

rather to the population as a whole, then this contradiction between full employment, income distribution, and environmental protection could be resolved. Although the implementation of a pollution control programme would still require that some portion of the productive resources of society be devoted to abatement rather than commodity production, the economic sacrifice of this commitment would not be borne by the members of a particular social class, in whose interest it would then be to oppose environmental protection.

A final issue which this analysis sheds light on is whether or not population growth is responsible for increased environmental pollution. The conclusion we have reached here is that maintenance of environmental quality is probably compatible with balanced expansion of gross output and population at annual percentage rates of $m + \tilde{n}$, and $\tilde{n}$ respectively. What this result reflects is that it is the waste residuals stemming from production and consumption, and not population size as such, which tend to lead to more environmental pollution and that this threat can be overcome if a sufficiently large portion of the capital stock and labour supply are devoted to waste treatment and recycling.

What this analysis also points out, however, is that a reduction in the rate of population growth might be helpful in establishing balanced expansion of the spaceship economy. Referring one last time to 6(90), we can see that when commodity production is relatively capital intensive, a lower population growth rate makes it more likely that the necessary condition for balanced growth can be satisfied at the existing wage-rental ratio. Finally, balanced expansion of the economy is even compatible with a zero rate of population growth since the effective labour supply continues to grow at the percentage rate of technical improvement, namely m percent annually. In conclusion, then, we should not expect that containing environmental pollution will dictate that economic growth cease in the foreseeable future. It will require, however, that the social relations of production be reconstituted so that full employment, distributive equity, capital accumulation and environmental protection are compatible with one another.

It would be interesting to explore whether and to what extent environmental pressure could "endogenize" technological change in the sense of Romer (1990), or could even endogenize an upswing/downswing in a long cycle of economic activity (Krelle, 1992). But such reasoning would reinforce our policy conclusions that under suitable economic designs, such as a spaceship economy, global environmental pollution need not retard growth, in fact may well accelerate it.

REFERENCES

Boulding, K.E., *Beyond Economics*, University of Michigan Press: Ann Arbor, MI, 1968.

Daly, H.E. and J.B. Cobb, *For the Common Good*, Beacon: Boston 1989.

Daly, H.E., "The Economic Growth Debate: What some Economists have learned but many have not", *Journal of Environmental Economics and Management* 14(4), 1987, 323-336.

d'Arge, R., "Essay on Economic Growth and Environmental Pollution", *Swedish Journal of Economics* 1971, pp 24-41.

Hung, V. T., Chang, P. and K. Blackburn, "Endogenous Growth, Environment and R & D", in Carraro, C. (ed.), Trade, Innovation and Environment, Kluwer Academic Publ.: Dordrecht, 1993.

Krelle, W., "A Theory on Long-Term Changes of Economic Growth", SFB 303, University of Bonn, Disc. Paper B-213, March 1992.

Mishan, E., "The Costs of Economic Growth", Praeger: New York 1967

Pezzey, J., "Economic Analysis of Sustainable Growth and Sustainable Development", The World Bank, Environment Department Working Paper No. 15, March 1989.

Romer, P.M., "Endogenous Technological Change", *Journal of Political Economy* 98, (5), 1990, S71-S102.

Solow, R.M., "A Contribution to the Theory of Economic Growth", *Quarterly Journal of Economics* 70, 1956, pp 65-94.

Stiglitz, J. and H. Uzawa (eds), "Introduction", *Readings in the Modern Theory of Economic Growth*, M.I.T. Press: Cambridge, Mass 1969.

Uzawa, H. "On a Two-Sector Model of Economic Growth, II" *Review of Economic Studies* 30, 1963, pp 105-118.

CHAPTER 8

OPTIMAL ECONOMIC GROWTH
WHEN CO_2 CONSTRAINTS ARE CRITICAL

8.1 INTRODUCTION

Past analyses of optimal economic growth behaviour have neglected the constraints of fossil fuel production and the accumulation of atmospheric CO_2 having a negative impact on production and productivity, as pursued in the more general context of Chapter 7.

The factors of production have been assumed to be either self regenerating (labour) or augmentable via production (capital) or technical progress. We explore the implications of accounting explicitly for the use of fossil fuel inputs (resources), constrained by a critical CO_2 budget. The optimal growth problem is developed here within the context of a familiar neoclassical, one-sector economy in which it is the objective to identify the path of capital accumulation that maximizes discounted per capita consumption over a finite planning horizon, subject to end point conditions on the stocks of capital and the critical CO_2 budget. This supplements the formulation of Chapter 2 in which a general model is developed for an infinite time horizon.

Such thoughts naturally relate to the problems of economic growth and sustainable development (Pearce, et al., 1989; Beltratti, 1996). Although it is beyond the purpose of this and the previous chapters to contribute to the discussion of these important issues, it is nonetheless expected that the present results support a reconciliation of economic growth and sustainability.

8.2 FORMULATION OF THE PROBLEM

For notational convenience we introduce the following definitions of variables and parameters.

$Y(t)$	= rate of output
$K(t)$	= capital stock
$L(t)$	= labour input
γ	= relative rate of technological progress in resource requirements
t	= time
$F(t)$	= stock of fossil fuel energy resource
$s(t)$	= savings ratio (fraction of total output devoted to capital investment)
ν	= relative rate of capital stock decay
$P(t)$	= population
μ	= social rate of time discounting
π_p	= relative rate of population growth
π_ℓ	= relative rate of labour force growth
k_o	= $K(0)/L(0)$ = initial capital-labour ratio
k_T	= $K(T)/L(T)$ = terminal capital-labour ratio
F_o	= initial stock of energy resources
F_T	= terminal stock of energy resources

The production function of the economy is assumed to be as follows:

$$Y(t) = \min[f(k(t), L(t)), e^{\alpha t} F(t)] \tag{1}$$

where f(.) is neoclassical and satisfies the Inada conditions.[1]

Equation (1) implies that if $f(K(t), L(t)) < e^{\gamma t} F(t)$, one may write

$$Y(t) = f(K(t), L(t)), \tag{2}$$

and

$$\dot{F}(t) = -e^{-\gamma t} f(K(t), L(t)). \tag{3}$$

(2) states that the rate of output is a function of the size of capital and labour inputs.

(3) states that the rate of CO_2 accumulation is proportional to the rate of output and that the

260

proportion diminishes as time passes due to exogenous technological advances that permit fossil fuel resources to be used more efficiently, or have them substituted by CO_2 benign energy technologies.

The savings-investment identity is given by (4):

$$s(t) \cdot f(K(t), L(t)) = \dot{K}(t) + vK(t).$$ (4)

The optimal growth problem is to find the time path for the savings ratio, s(t), $0 \le$ s(t) $\le$ 1, [2] that maximizes the following integral:

$$J = \int_O^T [1-s(t)] \frac{f(K(t), L(t))}{P(t)} e^{-\mu t} dt$$ (5)

which is simply the discounted sum of per capita consumption over the planning horizon [0, T].

If it is assumed that population grows at a given relative rate,

$$P(t) = P(0)^{\pi_p t}$$ (6)

and the input of labour according to a given relative rate,

$$L(t) = L(0) e^{\pi_l t}$$ (7)

then the optimal growth problem can be written simply as follows:

$$maximize \ J = \int_O^{\perp} (1-s) g(k) e^{-\rho t} dt$$ (8)

subject to

(a) $\dot{k} = sg(k) - \eta k$

(b) $\dot{z} = -g(k) e^{-\delta t}$

261

(c) $0 \le s \le 1$, $k \ge 0, z \ge 0$

(d) $e^{-\rho T} \cdot p(T) [F(T) - F_T] = 0, p(T) \ge 0, [F(T) - F_T] \ge 0$

(e) $e^{-\rho T} \cdot q(T) [F(T) - F_T] = 0, q(T) \ge 0, [F(T) - F_T] \ge 0$

(f) $k(0) = k_o$

(g) $F(0) = F_o$

Except where necessary the time arguments of the variables have been dropped. The function g(k) is simply

$$\frac{f(K, L)}{L} = f(\frac{K}{L}, 1) = f(k, 1) \qquad (9)$$

where k = K/L. The parameters ρ, η and δ are defined as follows:

$$\rho = -\pi_\ell + \pi_p + \mu \qquad (10)$$
$$\eta = v + \pi_\ell \qquad (11)$$
$$\delta = \gamma - \pi_\ell \qquad (12)$$

Assume $\rho > 0, \eta > 0$, and $\delta > 0$. P(0)/L(0) may be arbitrarily set equal to unity with no loss of generality. The variables, p(t) and q(t) are the undiscounted social prices of capital and resources, respectively, expressed in terms of consumption units.[3] It is necessary to place some further restrictions upon the end points in order to ensure that the problem as formulated above is interesting. For there to exist at least one path satisfying the end point conditions, it must be true that

$$\int_O^T g(k) e^{-\delta t} dt \le F_o - F_T \qquad (13)$$

for at least one savings trajectory, $\hat{s} = \hat{s}(t)$, for which it is true that

$$\int_O^T [\hat{s}g(k) - \eta k] dt \ge F_T - F_o \qquad (14)$$

262

That is, there must exist at least one savings trajectory that permits the economy to achieve the terminal capital-labour ratio without exceeding the limit on (critical) CO_2 accumulation. If only one path satisfies both (13) and (14) then it is perforce the optimum path. Although the problem is feasible if it is assumed that (13) and (14) are satisfied for at least one savings trajectory, the problem is of economic interest only if it is also true that

$$\int_0^T g(k) \, e^{-\delta t} dt > F_o - F_T \tag{15}$$

given the savings trajectory $s^* = s^*(t)$, where $s^*(t)$ is the optimal savings rate in the non-CO_2 constrained problem i.e. the problem where CO_2 accumulation of F is ignored.

8.3 DERIVATION OF THE NECESSARY CONDITIONS

Using the method of Pontryagin *et al* (1962), the necessary conditions for maximizing J are derived by forming the following Hamiltonian:

$$H = e^{-\rho t}\left\{(1-s)\,\dot{g}(k) + p[sg(k) - \eta k] + q[-g(k)\,e^{-\delta t}]\right\}. \tag{1}$$

To maximize J it is necessary first that H be maximized at each point in time with respect to s. This implies the following decision rules:

If $p > 1$, set $s = 1$.

If $p = 1$, set $s \in [0,1]$. $\qquad\qquad$ (2)

If $p < 1$, set $s = 0$.

It is also necessary that ρ and q satisfy the following differential equations:

$$\dot{p} = \rho p - \frac{\partial H e^{\rho t}}{\partial k} \tag{3}$$

$$\dot{q} = \rho q - \frac{\partial H e^{\rho t}}{\partial F} \tag{4}$$

Eliminating the partial derivatives, one obtains

$$\dot{p} = [(\rho+n) - sg'(k)]\,p - [(1-s)\,g'(k) - qg'(k)\,e^{-\delta t}] \tag{3'}$$

$$\dot{q} = \rho q \tag{4'}$$

Conditions (2), (3') and (4') in addition to those listed in 2(8) constitute the necessary conditions for a maximum and are the means of identifying the character of the optimal growth trajectory.[4]

8.4 ANALYSIS OF THE NECESSARY CONDITIONS

The task at hand is to reduce these conditions to a more tractable form. First note that differential equation 3(4') is easily solved:

$$q = q_o e^{pt} \tag{1}$$

Second, note that s can be effectively eliminated from differential equations (3') and 2(8a) using conditions (2):

$$\dot{p} = q_o e^{(p-\delta)t} g'(k) + \begin{cases} [p+\eta - g'(k)]p, & p>1 \\ [p+\eta - g'(k)], & p=1 \\ (p+\eta)p - g'(k), & p<1 \end{cases} \tag{2}$$

$$k = \begin{cases} g(k) - \eta k, & p>1 \\ sg(k) - \eta k, & p=1; \text{ all } s, \; 0 \leq s \leq 1 \\ -\eta k, & p<1 \end{cases} \tag{3}$$

Since (2) and (3) contain only k and p, except for the autonomous time terms, $q_o e^{(p-\delta)t}$ the necessary optimum behaviour for k and p can be deduced from a phase diagram constructed in (k, p) space.

The Case $\rho = \delta$

Assume temporarily that $\rho = \delta$. The time autonomous term is then reduced to a constant, q_o. Conditions 2(8e) and (1) require that $q_o \geq 0$. Consider first the case $0 \leq q_o < 1$. Then the $\dot{p} = 0$ stationary appears as in Figure 1, intersecting the horizontal line p = 1 at a non-negative value of $k, \tilde{k}$, such that

$$g'\tilde{k} = \frac{\rho + \eta}{1 - q_o} \tag{4}$$

The point k'' is determined by the solution to

$$g'(k'') = \rho + \eta \tag{5}$$

and indicates the asymptote toward which the $\dot{p} = 0$ stationary converges when p > 1.

265

Observe that $\dot{p} > 0$ at all points to the right of the stationary, and $\dot{p} < 0$ at all points to the left.

In the case $q_o \geq 1$, the $\dot{p} = 0$ stationary appears as in Figure 2. Observe that the horizontal line $p = 1$ is not intersected in the positive (k, p) quadrant except in the case $q_o = 1$, in which event intersection occurs at (0, 1). At all points below or to the right of the stationary $\dot{p} > 0$ and at all points above or to the left $\dot{p} < 0$.

Turning to the behaviour of k as indicated by (3), the $\dot{k} = 0$ stationary appears as shown in Figure 3. The hatched line represents the stationary. The solution to

$$g(k) / k = \eta \qquad (6)$$

yields k^{max}, the maximum sustainable capital-labour ratio. Note that $\dot{k} > 0$ at all points above or to the left of the stationary and that $\dot{k} < 0$ at all points below and to the right.

The $\dot{k} = 0$ and $\dot{p} = 0$ stationaries are combined in the two phase diagrams, Figures 4 and 5, corresponding to the cases $q_o \geq 1$ and $q_o < 1$ respectively.[5] Using the phase diagrams and the end point conditions 2(8d) through 2(8g) one can straightforwardly determine the optimal growth path.

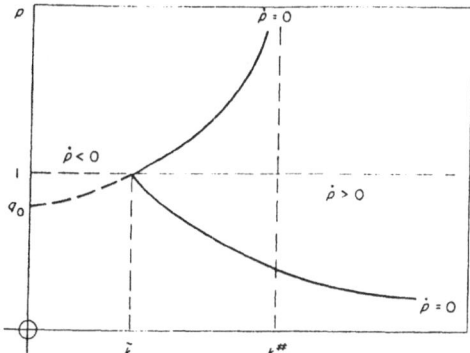

Figure 1.

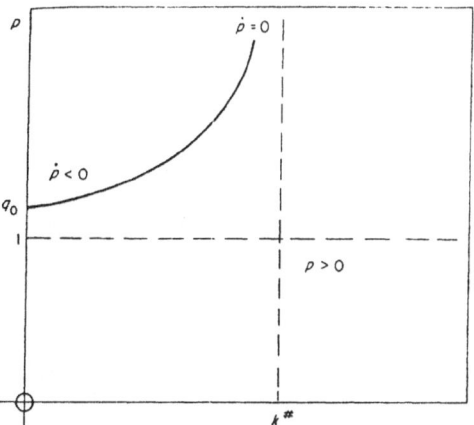

Figure 2.

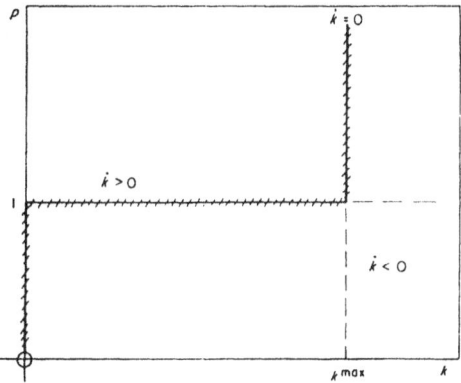

Figure 3.

267

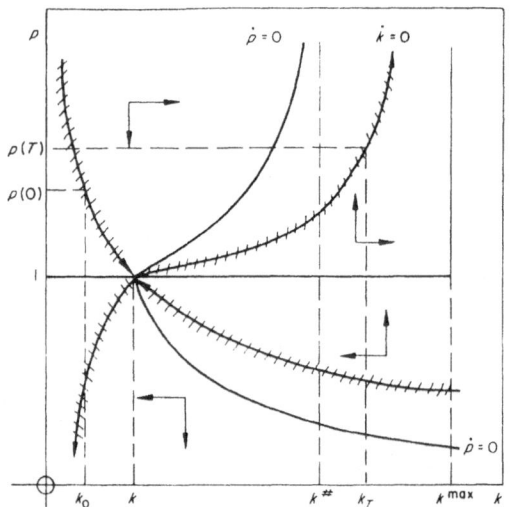

Figure 4.

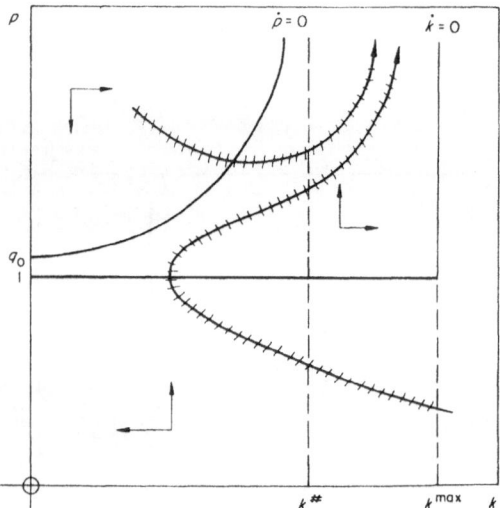

Figure 5.

268

The Unconstrained Case

In analogy to Shell, (1967), see also McKenzie (1986), we can show that in the case when the CO_2 constraint is not binding a typical optimal growth trajectory will be as shown in Figure 6. For $0 \leq t < t_1$ and $t_2 < t \leq T$, $s^* = 1$, and for $t_1 \leq t \leq t_2$, $s^* = \eta k^\# / g(k^\#)$. The corresponding phase diagram appears in Figure 7. Since it is assumed that

$$\int_0^T g(k(t)) e^{-\delta t} < F_o - F_T , \quad s = s*(t) \tag{7}$$

it must be true that $q_o = 0$ by condition 2(8e) and by (1). It follows that $\tilde{k} = k^\#$, so that the $\dot{q} = 0$ stationary becomes a vertical line at $k = k^\#$ for $p \geq 1$. Figure 7 shows the saddlepoint property of the point $(k^\#, 1)$ which Shell (1967) identified as the "modified Golden-Rule Turnpike", as an essential supplement of the basic Golden Rule concept [Phelps (1967)]. As the planning horizon gets longer, the proportion of time spent in the turnpike phase (t_1, t_2) grows larger, provided that t_1 and $T - t_2$ are finite as assumed. By definition,

$$g'(k^\#) = \rho + \eta \tag{8}$$

i.e. the rate of return on capital while on the turnpike equals the sum of the given relative rates of population growth, of capital decay, and of societal time discounting.

Fossil Fuel Constraint

When the fossil fuel constraint is binding, the trajectory shown in Figure 6 is no longer optimal since by assumption too much CO_2 will be accumulated by the end of the planning period. A typical optimal growth path in this new circumstance is illustrated in Figure 8. For $0 \leq t < t_1'$ and $t_2' < t \leq T$, $\tilde{s}$ (the "fossil fuel limited" optimum savings rate) is equal to unity. For $t_1' \leq t \leq t_2'$, $\tilde{s} = \eta \tilde{k} / g(\tilde{k})$, where

$$g' \tilde{k} = \frac{\rho + \eta}{1 - q_o}, \quad 0 < q_o < 1 \tag{9}$$

In the case $\rho = \delta$ and where feasibility is assured by conditions 2(13) and 2(14), q_o must be between zero and unity. This may be verified as follows: The time path of least energy use (L.E.U.) is one in which all investment is postponed until the last possible moment, ie for $t < t_s$, $s(t) = 0$ and for $t \geq t_s$, $s(t) = 1$. (See Figure 9)

Let k_s be the non-zero value of k at $t = t_s$. Note that (9) is such that there must exist a value of q_o between zero and unity, q_o^s such that $\tilde{k} = k_s$. If $F(T) = F_T$ at the end of the L.E.U. path, then this value of q_o is the optimal value. But if $F(T) > F_T$ at the end of the L.E.U. path, that path would satisfy the transversality conditions for an optimum only if q_o were equal to zero. In this case the L.E.U. path cannot be optimal. By inspection of the phase diagram shown in Figure 4 and of (9), it is apparent that $F(T)$ is a continuous function of q_o so that q_o may be set at some lower value, $\tilde{q}_o$, $0 < \tilde{q}_o < q_o^s < 1$ such that $F(T) = F_T$ and $\tilde{k} > k_s$. Selection of the appropriate value of $\tilde{q}_o$ will ensure that output along the growth path (and consequently energy consumption) will increase by just enough in comparison to the L.E.U. path to meet the terminal condition on energy stocks exactly.

The optimal growth path in the case discussed here includes a "turnpike phase", see Gottinger (1997). But this CO_2 emission limited turnpike differs from Shell's modified Golden-Rule turnpike in that the rate of return on capital is higher while the savings ratio and capital-labour ratio are both lower than on the modified Golden-Rule turnpike, due to the need to restrict fossil fuel use. This difference is directly deducible from comparison of (8) and (9).

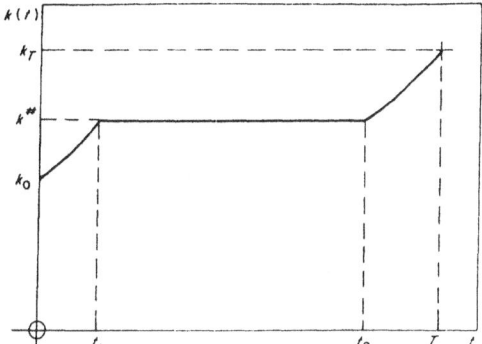

Figure 6.

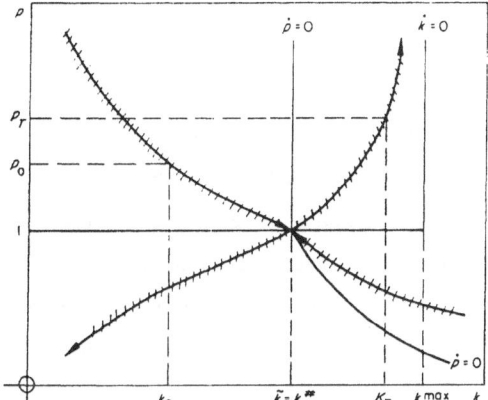

Figure 7.

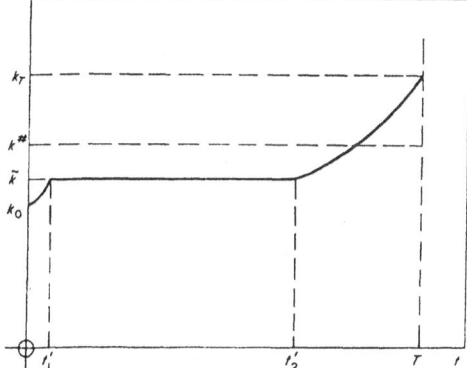

Figure 8.

271

Before turning to the case $\rho \neq \delta$, it will be useful to consider the meaning of the case $q_o \geq 1$, heretofore ignored. It was demonstrated above that $0 < q_o < 1$ for the class of problems that possess a feasible L.E.U. path. The case $q_o > 1$ corresponds to the class of problems not having a feasible L.E.U. path and thus possessing no optimal growth path. This can be motivated by the following illustration: If at $t = 0$, $q_o > 1$, then the social value of energy resources consumed in producing a unit of output will exceed the social value of the output itself unless it is also true that the social value of a unit of investment, $p(0)$, exceeds q_o. In this latter event it will be socially desirable to engage in production since the social value of output (allocated entirely to investment according to the necessary conditions) will exceed the social value of energy resources consumed in production $(p(0) > q_o)$. But the necessary conditions for an optimum indicate that if $q_o \geq 1$ and $p(0) > q_o$, s must equal unity throughout the growth trajectory. This is not possible since by assumption such a trajectory would violate the fossil fuel use constraint. If it is assumed that $p(0) < 1 \leq q_o$, then the way to maximize social welfare is to cease production entirely, since energy resources are socially more valuable *qua* resources than embodied in final output. This anomaly is due to the fact if production is not reduced to zero for a finite period of time then the fossil fuel use constraint must be exceeded, even on the L.E.U. path.[6] The borderline case $q_o = 1$ can be interpreted as a limiting case in the set of problems where the L.E.U. path is the only feasible path and is thus also the optimum path. As k_s of the L.E.U. path $\rightarrow 0$, then $q_o \rightarrow 1$.

The Case $\rho \neq \delta$

Turning to the more general case $\rho \neq \delta$, the differential equation for the social price of capital, 2(14) will contain an explicit time dependent term, $q_o e^{(\rho-\delta)t}$. As a result, the $\dot{p} = 0$ "stationary" as projected onto the (k, p) phase plane will shift systematically with the passage of time. Suppose that the phase diagram appears as in Figure 4 at $t = 0$. If $\rho < \delta$, then the influence of the time dependent term in 2(14) will diminish as time passes. The point $\bar{k}$ will approach the point $k^{\#}$ asymptotically and the appearance of the phase diagram will come to resemble that of Figure 7, constructed for the case $q_o = 0$. In this example, technological progress in fossil fuel use proceeds at such a rapid pace, relative to the social rate of time preference and the rate of population growth, that the fossil fuel use constraint virtually ceases to be binding. Optimal growth models that have ignored CO_2 emission limits

272

have rested, at least tacitly, upon such an assumption. If $\rho > \delta$, then the time term increases in importance as time passes, and $\bar{k}$ will move toward the origin. Eventually, $q_o e^{(\rho-\delta)t}$ will equal and then exceed unity. The phase diagram will then appear as in Figure 5, with the $k = 0$ intercept of the $\dot{q} = 0$ "stationary" curve rising steadily.

Despite the fact that the case $\rho > \delta$ leads eventually to a phase that corresponded to an infeasible problem in the $\rho = \delta$ case, it is not true that all cases where $\rho > \delta$ are infeasible. If a feasible L.E.U. path exists, then an optimal solution may exist, depending upon the existence of a feasible savings ratio. Given any feasible L.E.U. trajectory (see Figure 9), it is possible to set q_o at such a value, q_o^s , lying between zero and unity, that

$$g'(k_s) = \frac{\rho + \eta}{1 - q_o^s e^{(\rho-\delta)t_s}} \tag{10}$$

If q_o^s is so chosen, then it must be true that $\bar{k}(t_s) = k_s$ since by definition

$$g'(k(t)) = \frac{\rho + \eta}{1 - q_o e^{(\rho-\delta)t}} \tag{11}$$

Thus, q_o can be chosen so that at time t_s, $\bar{k}(t_s)$ will coincide with the value of $k(t_s)$, where

$$k(t) = k_o e^{-\eta t} \tag{12}$$

the latter being the capital growth equation for the phase s = 0. This point is illustrated by example in Figure 10, which shows the curve $\bar{k} = \bar{k}(t; q_o^s)$ intersecting the L.E.U. path at t = t_s.

If the L.E.U. path is the only feasible path, then q_o^s will be the optimal initial value of q. The corresponding phase diagram is illustrated in Figure 11. The hatched line with arrows marks the optimal trajectory. The position of the $\dot{p} = 0$ stationary is shown at both t = 0 and t = t_s. It is readily apparent that if F(T) = F_T, all the necessary conditions are satisfied. But if F(T) > F_T on the L.E.U. path, transversality condition 2(8e) is not satisfied, since

$$q(T)\,[F(T) - F_T] \neq 0$$

In this event the L.E.U. path cannot be optimal.

When the L.E.U. path is not optimal, it is necessary to verify the feasibility of the savings ratio in order to demonstrate satisfaction of the necessary conditions and consequently to characterize the optimal trajectory. Consider the case where $k_o = \tilde{k}(0, \tilde{q}_o)$ and $k_T = \tilde{k}(T, \tilde{q}_o)$. Assume that $F(T) = F_T$, then the necessary conditions require that the savings ratio, s, be set such that $k_t = \tilde{k}(t, \tilde{q}_o)$. The problem is that this required savings ratio, $\tilde{s} = \tilde{s}(t)$, may violate the condition $0 \leq \tilde{s}(t) \leq 1$.

From (11) one can show that

$$\dot{\tilde{k}} = \frac{(\rho-\delta)\,\tilde{k}}{e(\tilde{k})} \cdot [1 - \frac{f'(\tilde{k})}{p+\eta}] \quad where \quad e(\tilde{k}) = -\frac{f''(\tilde{k})\,\tilde{k}}{f'(\tilde{k})} . \tag{12}$$

But 2(8a) requires that

$$\dot{k} = \tilde{s} f(k) - \eta k$$

Setting $\dot{k} = \dot{\tilde{k}}$ according to assumption and solving for $\tilde{s}$, one obtains

$$\tilde{s} = \left\{ \frac{\rho-\delta}{e(\tilde{k})} [1 - \frac{g'(\tilde{k})}{\rho+\eta}] + \eta \right\} \frac{\tilde{k}}{f(\tilde{k})} \tag{13}$$

It is easily shown that this condition could be violated along the trajectory $\tilde{k} = \tilde{k}(t, \tilde{q}_o)$.[7]

Since the case at hand assumes $\rho > \delta$ and since $\tilde{k} < k^{\#}$ by necessity, only the lower bound on $\tilde{s}$ is of interest, by (32), which shows that $\dot{\tilde{k}} < 0$. The question is whether

$$\left| \dot{\tilde{k}} \right| > \eta \tilde{k}$$

at any point along the trajectory. If not, then s can always be set equal to the desired value, and the optimal time path of k is indicated by the hatched line shown in Figure 12.

It can be shown, as in the $\rho = \delta$ case, that if a feasible L.E.U. path exists, then an optimal path exists, provided that a feasible savings ratio can be found. Figure 13 again displays the L.E.U. path and the $\tilde{k}$ trajectory for which q_o is chosen so that

$$k_s = \tilde{k}(t_s; q_o^s)$$

As noted above, if $F(T) > F_T$ on the L.E.U. path, then q_o (and by implication, s) has not been chosen optimally. Also contained in the figure is an alternative path for which q_o has been set at the optimal level, $\tilde{q}_o < q_o^s$. Since the level of fossil fuel use along the growth path is a continuous function of q_o, one can always find a $\tilde{q}_o$ that is low enough to ensure that the terminal fossil fuel stock satisfies the condition

$$F(T) = F_T$$

thus satisfying all the necessary conditions. For $t < t_1$ and $t_2 < t \leq T$, the optimum savings ratio is set at unity. For $t_1 < t \leq t_2$, $\tilde{s}$ satisfies (32). During the intermediate phase, the optimal growth characteristics of the CO_2 emission constrained economy compare to those of an economy on a modified Golden-Rule turnpike in the same qualitative way as in the case $\rho = \delta$: The savings ratio and capital-labour ratio are both lower than in the unconstrained economy. However, contrary to the $\rho = \delta$ case, both are declining as $t \rightarrow t_2$.[8]

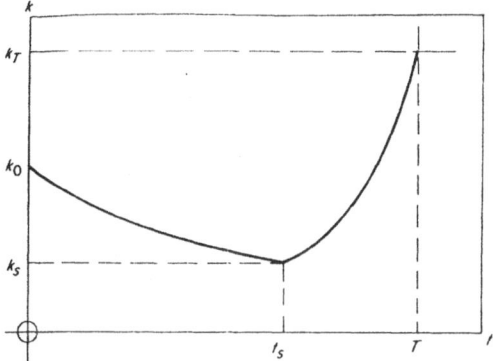

Figure 9.

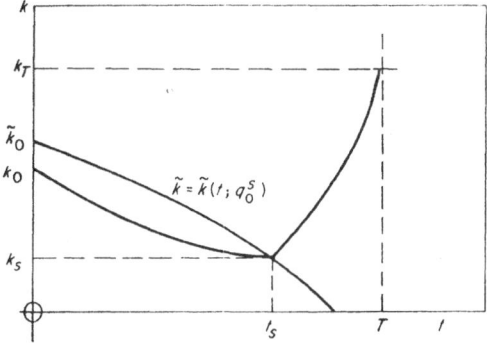

Figure 10.

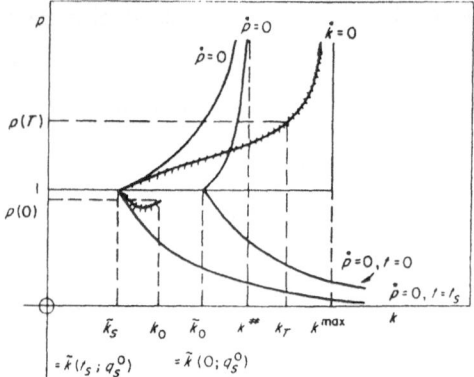

Figure 11.

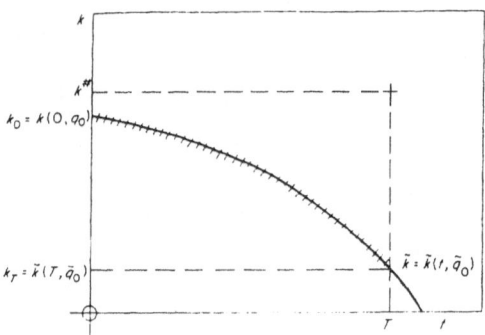

Figure 12.

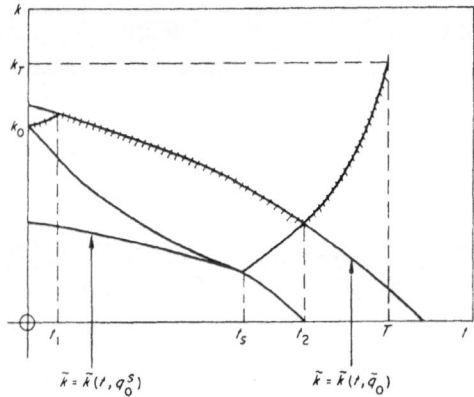

Figure 13.

276

8.5 CONCLUSION

The analysis shows that when consumption of fossil fuel resources under CO_2 emission limits is incorporated into an optimal economic growth model, the result is a tendency to postpone capital accumulation and to spend time on turnpike growth paths where capital is used less intensively than in a non-CO_2-constrained model of the economy. In the extreme case, where the only feasible growth path is the path of minimum fossil fuel use, all capital accumulation is postponed until the last possible moment.

Of course, if technical progress comes with capital formation only such a result would have to be conditional on the impacts of technical change (Chapter 4).

1. If f(.) is neoclassical and satisfies the Inada conditions (see Hamberg, 1971) it is twice continuously differentiable and has the following other properties:

$$\lambda f(K,L) = f(\lambda K, \lambda L), \partial f/\partial K > 0, \partial f/\partial L > 0, \partial^2 f/\partial K^2 < 0, \partial^2 f/\partial L^2 < 0,$$

$$\partial^2 f/\partial K \partial L > 0, \lim_{(K/L) \to 0} f(K/L, 1) = 0, \lim_{(K/L) \to \infty} f(K/L, 1) = \infty,$$

$$\lim_{(K/L) \to 0} \frac{\partial f(K/L, 1)}{\partial (K/L)} = \infty, \qquad \lim_{(K/L) \to \infty} \frac{\partial f(K/L, 1)}{\partial (K/L)} = 0.$$

2. Although it is true that situations where s = 1 or s = 0 are hardly likely to occur in the real world, the optimal growth problem thus formulated differs little qualitatively from one in which it is assumed that $0 < s_{min} \le s \le s_{max} < 1$ and is less cumbersome to deal with.

3. The terms $p(t)^{-\rho t}$ and $q(t)e^{-\rho t}$ are actually undetermined Hamiltonian multipliers, analogous to Lagrangian multipliers in static problems. See Pontryagin (1962), especially Chapter I, pp 1-74.

4. For reasons of space we present no proof of the sufficiency conditions for a maximum. Such a proof would follow the lines of that employed for a similar problem by K. Shell (1967), pp. 11-13. Assuming that maximization of the Hamiltonian, H, at each instant, offers no difficulty, then concavity of H in the state variables (k and F) assures sufficiency. In this problem, H is concave in k and F if the necessary conditions are satisfied.

5. By necessity, $k^{max} > k^{\#}$ in Figures 4 and 5 due to the assumed characteristics g(k). Figure 4 is drawn so as to illustrate the saddlepoint property of the point $(\tilde{k}, 1)$, which possesses unique convergent and divergent arms. This property follows from the Lipschitzian character of the right hand sides of the differential equations for p and k in the regimes p > 1 and p < 1. See Shell, (1967) especially pp. 5-6.

6. The optimal growth problem, as formulated here, implies that capital and labour are always fully employed if production is non-zero. In a more complicated model that permitted unemployment of capital and/or labour, there would be feasible growth paths that implied less fossil fuel use than the L.E.U. path of the problem at hand.

7. Suppose a Cobb-Douglas production function obtains, ie

$$\tilde{k} = \tilde{k}^{1-\beta}, \beta = constant, \ 0 < \beta < 1 \ then$$
$$\tilde{s} = \left\{ \frac{\rho - \delta}{\beta} [1 - \frac{\tilde{k}^{-\beta}(1-\beta)}{\rho + \eta}] + \eta \right\} \tilde{k}^{\beta}$$

It is easily seen that $\tilde{k}$ may take on values that imply infeasible values for $\tilde{s}$.

8. In the $\rho < \delta$ case they are, of course, both rising asymptotically toward their modified Golden-Rule turnpike values.

REFERENCES

Beltratti, A., Chichilnisky, G. and G. Heal, "Sustainable Growth and the Green Golden Rule", in Sustainable Economic Development: Domestic and International Policy, Nota di Lavoro 61.93, Fondazione ENI Enrico Mattei, Milan, 1993.

Beltratti, A., Models of Economic Growth with Environmental Assets, Kluwer Academic Publ.: Dordrecht, 1996

Gottinger, H. W., "Polluted and Sustainable Turnpikes", Japanese Journal of Business and Economics, 77, 1997, pp. 47-74.

Hamberg, D., *Models of Economic Growth*, Harper & Row, New York 1971, Chapter 2.

McKenzie, L. W., 'Optimal Economic Growth, Turnpike Theorems and Comparative Dynamics', Chapter 26 in Arrow, K. J. and M. Intriligator (eds) *Handbook of Mathematical Economics*, Vol III, North Holland: Amsterdam, 1986, pp 1281-1355.

Pearce, D., Markandya, A. and E. B. Barbier, *Blueprint for a Green Economy*, Earthscan Publishing, London, 1989.

Phelps, E. S., Golden Rules of Economic Growth, North Holland: - Amsterdam 1967, Chapters 1 and 4.

Pontryagin, L. S., *et al., The Mathematical Theory of Optimal Processes*, Interscience Publishers, New York and London, 1962.

Shell, K., ed., *Essays on the Theory of Optimal Economic Growth*, The M.I.T. Press, Cambridge, Massachusetts, 1967.

CHAPTER 9

UNCERTAINTY, VALUE OF INFORMATION
AND
GREENHOUSE GAS EMISSIONS

9.1 INTRODUCTION

The concentration of greenhouse gases in the atmosphere could be perceived as a non degradable stock pollutant which causes unknown environmental costs. In view of designing commonly agreed (globally negotiated) regulatory strategies against a perceived but on its scale unknown threat the problem is one of preserving regulatory options, and placing values on the preservation of such options before irreversible but uncertain damage can occur.

We explore an optimal regulatory regime in which regulating a non-degradable pollution stock, e.g. the accumulation of greenhouse gases in the atmosphere, would serve two purposes, as found in a more general context of resource economics by Conrad (1992):

"First, if environmental damage is unexpectedly higher in the future the level of 'regret' will be lower. Second, if environmental damage is unexpectedly lower, a low pollution stock today preserves the option of increasing the stock in the future".

Such a design relates to problems of optimal stopping.

Choosing the level of GHG emission-limiting regulations that will maximize social welfare by optimally balancing the costs of emission control against the benefits of decreased environmental damage is inherently not possible, because of pervasive uncertainty about the likely size of the critical GHG budget, its relationship to the quantity of GHG emitted, the effects of GHG in the atmosphere, and the appropriate valuation of these consequences. (Lave, 1991). Moreover, we can expect to learn more about each of these areas of uncertainty through continuing scientific-technological research, and through observation of atmospheric responses to past and current GHG emissions (Flohn, 1990). Because overall we can expect these uncertainties to diminish over time, the appropriate policy is likely to be an incremental and dynamic one. The risk to delaying before further restricting GHG emissions is that, if significant emission reductions become necessary to prevent serious adverse consequences, their cost may be much larger than if emission reductions begin sooner.

On the other hand, the risk to adopting further restrictions now is that these restrictions may later prove to have been unnecessary; the costs incurred would have produced no benefits. The question is analogous to that of whether to purchase insurance; as recently formulated by Manne and Richels (1992); by imposing additional regulations now, we incur immediate costs in exchange for a potential reduction in the costs of preventing and adapting to future GHG accumulation. More recently, alternative approaches have been

280

followed by Peck and Teisberg (1993), Parry (1993), Hammitt et al. (1992), Kolstad (1992), and Lempert et al. (1994).

This chapter attempts to provide insight to this question. In order to better motivate insight into the structural features of making decisions under imperfect information, we first give an infinitive, illustrative explanation in Section 2. We develop a general formulation of the policy question which can be conceived as an infinite horizon stochastic dynamic program with learning. (Bertsekas, 1976, Part II). This formulation clarifies the issues, but is mathematically hard to cope with. To provide more explicit guidance, we take advantage of specific features of this problem to develop a simplified decision framework. Because of the long time delay in the relationships between GHG emissions, accumulation and effects in the atmosphere, we can structure the policy choice so that the environmental damages and benefits are approximately the same under each policy. Thus, the framework focuses attention on a comparison of the expected economic costs of alternative regulatory strategies.

The simplified framework characterizes the degree of belief about the severity of the GHG problem such that the expected economic cost of awaiting future scientific revelations before deciding at what level to regulate is greater or smaller than the expected cost of imposing interim regulations now. This required 'degree of belief' (subjective probability) defines the confidence policy makers should have in the proposition that without further emission restrictions, GHG accumulation will occur and produce substantial adverse effects.

Because alternative regulatory strategies considered in this framework are constructed to produce equivalent GHG concentration levels, it is not necessary to estimate the relationship between reduced GHG concentration and avoided damages, nor to value these damages.

This feature considerably reduces the information required to reach a decision and simplifies the analysis, since the economic costs of alternative control strategies may be compared directly. An important limitation of the analysis, however, is that it ignores the possibility that we may not learn whether GHG accumulation is likely to produce severe consequences until it is too late to prevent it.

Other formulations of uncertainty resolving regulatory strategies are possible (Kelly and Kolstadt, 1996) and they are related to the dynamic framework considered here.

281

9.2 AN ILLUSTRATIVE EXAMPLE - EVALUATING A CLIMATE RESEARCH AND MONITORING PROGRAMME (CRMP)

The Expected Value of Perfect Information (EVPI) can be put to test in the context of the initial analysis of policy options.

The value of improved information by a climate research and monitoring programme (CRMP) is of strategic consideration for identifying uncertainty in policy analysis. It is supposed that the policy-maker is interested in minimizing the expected damage over time generated by a persistent change of global climate with all its consequences.

The CRMP, consisting of a set of diagnostic tests, conducted in parallel or sequentially is considered to be an information generating device for finding out whether there is a significant climatic change with disruption impacts (A) or only a slight/moderate change with adaptable effects (B). Tests are used here in a general sense, involving either laboratory tests, physical findings, observations, experiments, and other scientific knowledge.

We all know that some information generating services are imperfect, ie error bound, so that we are left asking questions about the reliability and predictive accuracy of the test(s).

For this purpose we could set up a Test-Reliability Matrix for our simple problem.

Test Results	Climatic State	
	A	B
A	True Positive (TP)	False Positive (FP)
B	False Negative (FN)	True Negative (TN)

Table 1: Test-Reliability Matrix of Outcomes

Here the entries in this table, read along the first or second row respectively, could be interpreted as:

First Row:

"Test results indicate global climate to be in State A and State A is true (TP)".
"Test results indicate global climate to be in State A and State A is false (FP)".

Second Row:

"Test results indicate State B and State A is true (FN)".
"Test results indicate State B and State B is true (TN)".

Since by the nature of the problem State A is far more serious than B, ensuring significant higher costs (or loss of welfare) of human adaptation, it therefore requires immediate action by the policy maker. The Test-Reliability Matrix reflects this in the naming of the entries. (If the case A has been ruled out, one could set up another table comparing B with C, etc, so that the climatic state with the highest priority, requiring most policy attention, is taken proper care of). Of particular concern here are those policies which on the basis of the test results are implemented for the false climatic state (false positive) and those for which the results miss the true state (false negative). Suppose then, on the basis of these tests, for climatic states A and B, completely different policies D_1 and D_2 are suggested, for instance D_1 may involve a policy of severely constraining global fossil fuel use, D_2 is a low cost adaptation strategy. Suppose further we have different observational evidence on policies D_1 and D_2 with regard to impact costs (measured by some cost severity index 0 to 100), we could set up an outcome table on the decisions (costs) of policies D_1 and D_2.

Policies	Climatic State	
	A	B
D_1 (Act)	15	7
D_2 (Wait)	30	0

Table 2: Outcome of Decisions (Costs) Measured by Some Cost Severity Index

The numbers in the entries are used here only for illustrative purposes. It might be advisable to decompose the data according to country or regional groupings that could reveal quite distinct cost severity indices, thus the overall aggregate average cost matrix may not be applicable to group specific circumstances. (For collecting enough group specific, disaggregate data we may run into difficulties of sufficient data acquisition. In that case we may apply advanced statistical techniques, ie multiple regression analysis, for overcoming these difficulties.)

For any policy, D_1 and D_2, the average impact cost, given as the expected value, can be computed after specifying the probability of each climatic state, A or B.

Unless one has a sufficient database one often finds it difficult to calculate the probability of the climatic state.

In this case the policy makers are required to make a reasonable, balanced judgement based on the information available and come out with a subjective probability reflecting their professional judgement. Various methods to attain a subjective probability can be applied, De Finetti (1972), Fishburn (1982), in terms of betting quotients, or comparing climatic states with events for which well known (objective) probabilities exist.

Suppose the policy maker's prior probability of state A is $P(A) = 0.05$, and for B it is $P(B) = 0.95$. Then, on the basis of Table 2, we compute the expected value of policy:

$$EV(D_1) = 0.05 \times 15 + 0.95 \times 7 \quad = 0.75 + 6.65$$
$$= 7.40$$

$$EV(D_2) = 0.05 \times 30 + 0.95 \times 0 \quad = 1.50$$

Clearly, the action with the lowest average impact cost is best. A correct policy always has lower costs. These simple computations show that if you want to make a *terminal* decision or the equivalent, and the costs of gaining information about specifying P exceed the benefit of this information, for policies with 0.05 probability of having "high impact" cost state A, it is better to apply D_2 than D_1. In fact, looking at the average impact costs, D_1 costs 7.40 (impact) units, D_2 costs only 1.50 units. This is so because the consequence of applying D_2 if A is true is not severe enough to outweigh its small probability. (The underlying assumption is that waiting a couple of years, not necessarily decades, substantially increases the impact cost of climatic state A, but does not make the cost *prohibitive*. If the cost is prohibitive, no probability, however small, would be able to compensate for it).

As we can summarize, at this point, the best decision depends on the probabilities of the various climatic states and on the impact costs associated to the given status. By fixing the impact costs, one can easily determine the threshold probability at which point it becomes advisable to switch from strategy D_1 to D_2. (Analogously, by fixing the probabilities one can also vary the impact costs to the threshold point where both policies are equivalent in terms of expected costs.)

The threshold probability can be calculated as follows:

$$EV(D_1) = P \times 15 + (1 - P) \times 7 \quad = 7 + 8P$$

$$EV(D_2) = P \times 30 + (1 - P) \times 0 \quad = 30P$$
Equalizing, $\quad EV(D_1) = 7 + 8P = EV(D_2) = 30P$
yields $P = 0.31$ (cut-off point)

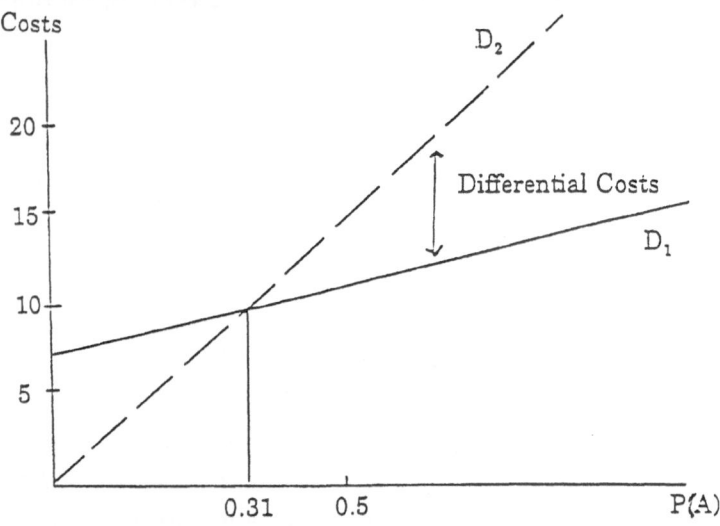

Figure 1: Impact Costs Depending on Threshold Probabilities

One can simply interpret Figure 1 along the following lines.

We graph the impact costs of D_2 (dashed line) and D_1 (solid line) as a function of the probability of A. If there is no chance of A occurring, there is no cost to D_2, but there is always a cost to the more radical "act now" policy D_1, ie the impact cost of the policy with an outcome "false positive". To the extent that A becomes more likely, the "D_2-strategy" quickly becomes more costly. Up to the cut-off point one should adopt a "wait-strategy", above this level one should switch to a more radical policy.

If we consider that a climate research and monitoring programme consists of a sequence of tests to be performed, it appears commonsense reasoning that under these circumstances a CRMP should correctly classify these climatic states that turn out to be true states. Since this requirement can only be fulfilled under exceptional, ideal circumstances, we could consider this state as our reference system and set out to enquire about the costs that obtain in such a system. It is clear that these costs cannot be decreased given the present level of research and monitoring technology.

286

In a perfect programme there are no false positives and no false negatives. Since the value of a perfect CRMP is one that predicts a state correctly and consequently leads to an appropriate policy, one would naturally arrive at the value of information that minimizes predictive mistakes emanating from any CRMP, coming close to a perfect programme. In other words by improving the predictive policy situation one asks how much is more information worth?

By referring to Table 2, a perfect CRMP yields impact costs by computing:

$$EV(D_1 \mid perfect) \quad = P \times 15 + (1 - P) \times 0 \quad =$$
$$0.05 \times 15 + 0.95 \times 0 \quad = 0.75$$

as compared to:

$$EV(D_1 \mid imperfect) = 0.05 \times 15 + 0.95 \times 7 \quad = 7.40$$

In verbal interpretation, applying D_1 in case of a perfect CRMP has a minimal impact of 0.75, an unavoidable cost, whereas applying D_1 given imperfect information has a cost 7.40. This means that the cost of action in the light of perfect information is roughly ten times less than the cost associated to the best action on undifferentiated states. The expected value of perfect information (EVPI), therefore, is equal to the difference of these two, since the impact cost of 0.75 appears unavoidable unless better climatic engineering (technologies) is available.

To emphasize the point of optional policies with perfect and imperfect information, we shall refer to costs for each state in terms of regrets caused by mistakes (to set up a regret matrix is a familiar procedure in statistical decision theory).

Policies	Climatic State	
	A	B
D_1	0	7
D_2	15	0

Table 3: **Outcome Expressed as Differential Impact Costs Due to Improper Policies**

The number 15 in the lower left entry of the matrix comes from 30, the costs of a "wait strategy" minus 15, the costs of the only correct policy, D_1. The upper left hand entry is zero, since the action is correct. The upper right hand entry remains as it is since the cost of the correct action, D_2, to be deducted, is zero.

Up to this point we considered only the value of perfect information as compared to imperfect information, taking the unavoidable impact costs of a correct policy as a basic reference point. The situation where perfect information can be acquired is rare, whereas partial information is often obtainable. (Nevertheless, the expected value of perfect information is useful because it provides an upper bound to that for partial information.) A related concept is the expected value of including uncertainty (EVIU) when perfect information is not available but instead some random variable giving imperfect (partial) information (see Henrion, 1990, Chap. 12).

More generally, we could exhaust the whole spectrum on evaluating different CRMPs and calculating the value of information in terms of impact costs. A crucial point in a comparative evaluation of these programmes is the validity of their test results, that is the degree of accuracy according to which these tests identify and classify the correct climatic states.

Let us consider only a possible result of test validation for illustration purposes.

Test Results	Climatic State	
	A	B
A	0.8	0.1
B	0.2	0.9

Table 4: Test Validations

We assume such numbers can be obtained indicating the reliability of such tests.

How do test results affect estimates of the probability of climatic change? Bayesian statistics provide techniques for revising initial or prior probabilities in the light of new information. The information must be new to have any effect.

In statistical analysis, the events are said to be independent if information about the occurrence of one event does not change the probability that the other event occurred.

Independence of climatic evidence is hard to judge. To estimate whether two tests are really independent in their predictions, one may have to collect substantially more cases. Sometimes there may be theoretical reasons to believe tests are independent; one may involve chemical, the other physical methods. Bayes' theorem weights prior probabilities by their likelihood; it follows from the definition of conditional probability.

The conditional probability of A given that the test indicates A is defined by:

$$P(A \mid Test = A) \quad = \quad P(A \text{ and } Test = A)/P(Test = A)$$

i.e. it is defined to be the probability of both A and a true A-test result divided by the total probability of a positive A test result. If, as in our example, the prior probability, $P(A)$, is 0.05, the test identifies 80 per cent of A correctly, so $0.05 \times 0.8 = 0.04$ is the probability being identified correctly as having A by the test. In the denominator is the 10 per cent of B, falsely called A by the test.

The probability of the test saying A is 0.05 x 0.8 + 0.95 x 0.1 = 0.135, and that 0.04/0.135 = 0.296 is the probability of A being correctly identified.

A similar calculation shows:

$$P(A \mid Test = B) = \frac{P(A \text{ and } Test = B)}{P(Test = B)}$$

$$= \frac{0.05 \times 0.2}{0.05 \times 0.2 + 0.95 \times 0.9} = 0.012$$

that in only 0.012 of the circumstances for which the test says B we will have A.

Let us see how we could assemble the various bits of information contained in Bayes' theorem: the differential cost (regret) matrix and the test validity data, forming the likelihood, to construct the value of a test represented only by a single criterion, the impact cost.

In general, the value of the test is simply calculated by subtracting the average costs of the chosen action before the information of the test is available from the chosen action afterwards.

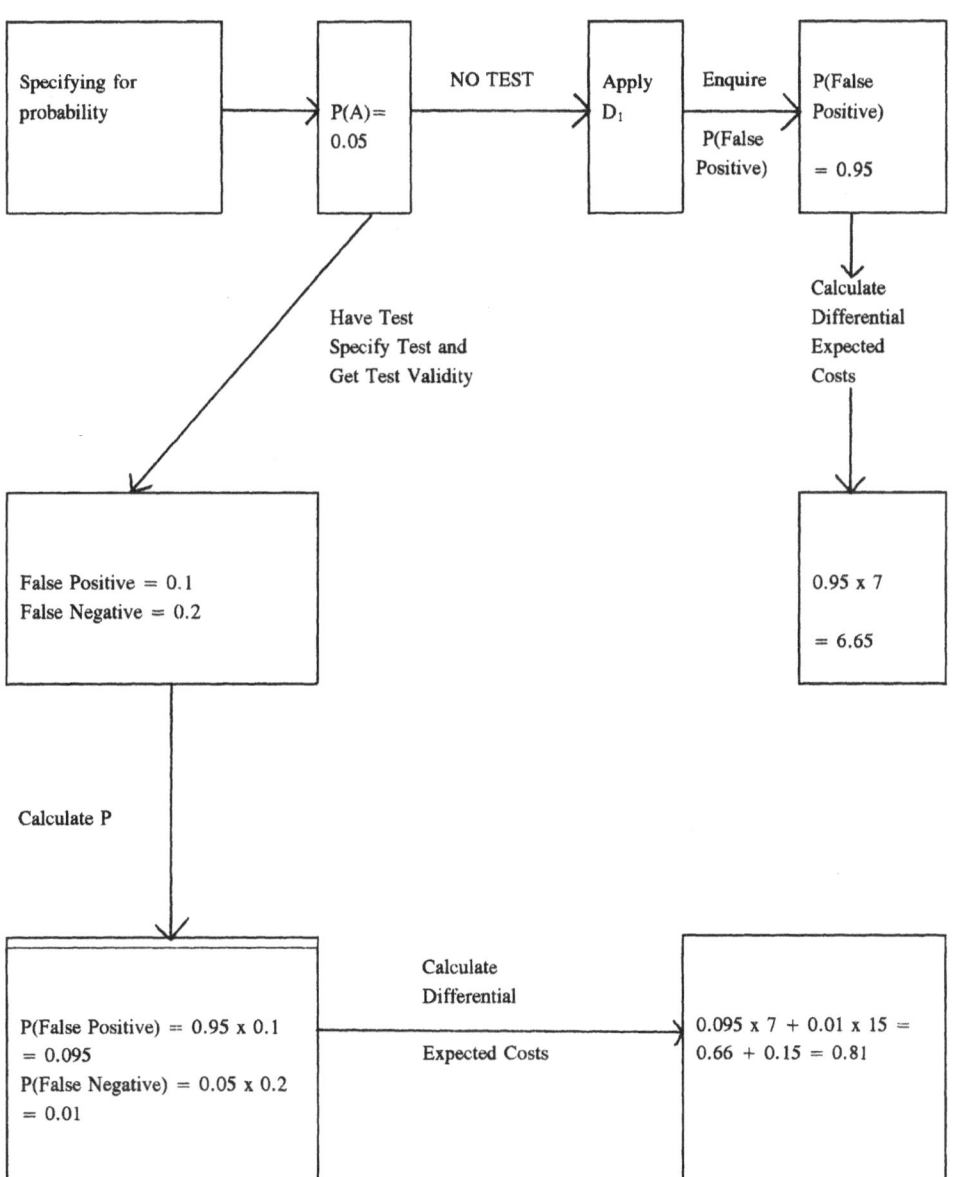

Specifying for probability	→	$P(A) =$ 0.05		NO TEST →	Apply D_1	Enquire	P(False Positive)	

NO TEST

P(False Positive)

Enquire

P(False Positive) = 0.95

Have Test
Specify Test and
Get Test Validity

Calculate
Differential
Expected
Costs

False Positive = 0.1
False Negative = 0.2

0.95 x 7

= 6.65

Calculate P

Calculate
Differential

P(False Positive) = 0.95 x 0.1
= 0.095
P(False Negative) = 0.05 x 0.2
= 0.01

Expected Costs

0.095 x 7 + 0.01 x 15 =
0.66 + 0.15 = 0.81

Table 5: Flow Chart to Calculate the Value of a Test

These calculations presented in a flow chart form use the added costs of mistakes over correct actions in the differential cost matrix. Applying D_1 immediately creates costs of 7 in state B. Given the prior probability on B additional cost without the test amounts to 0.95 x 7 = 6.65.

Applying D_1 and conducting the test, yields an expected differential cost of only 0.81. Thus the test saves costs of 5.84 = 6.65 - 0.81. Thus, in this context it is highly justified by using reduction of impact costs as a single criterion. In principle, the same calculations go through in a sequence of tests making the entire CRMP.

9.3   THE VALUE OF INFORMATION IN A STOCHASTIC DYNAMIC PROGRAM

This section formalizes the policy problem of limiting GHG emissions to prevent or limit global warming as a stochastic dynamic program with learning. We sketch the general problem and develop a simplified decision framework, based on principles set out in Sections 2 and 3, in which the value of information is derived as a solution to the stochastic dynamic program.

The policy problem of whether and when to restrict GHG emissions can be formulated as a stochastic dynamic program with learning over an indefinite, perhaps infinite, horizon. Define the following notation: t, discrete measure of time (for example, one unit = one year). t increases with calendar time.

s_t, the state variable, describes the atmospheric burden of GHGs at time t. s_t is a complex function of historical emissions of GHGs from anthropogenic and other sources, past solar activity, and possibly other factors. It may be adequate to approximate it by $s_t = \Sigma_l\, a_s(l)\, e_{t-l}$, that is, as a weighted sum of past GHG emissions, where the weights depend on the GHGs, atmospheric lifetimes, relative depletion efficiencies, and possibly on other factors.

θ, the state of nature, characterizes how damaging GHGs are. θ includes the relationships between GHGs and global warming, between global warming and increased solar radiation, and between increased solar radiation and damages in the biosphere.

$D(s_t,y_{dt},\theta)$, current damage from global warming. D depends on the state s_t, certain aspects of current technology y_{dt} (e.g. mitigating technologies), and the state of nature θ.

p_t, current GHG production, assumed equal to current use.

e_t, current GHG emissions. $e_t = \Sigma_l\, a_e(l; y_{et})\, p_{t-l}$, that is, e_t is a weighted sum of past production where the weights describe the time path of emissions from each GHG application. These weights depend on those aspects of current technology y_{et} that affect the timing of emissions from specific applications and the share of GHGs produced that are eventually emitted.

$B(p_t;y_{bt})$, economic benefits from current GHG use p_t. The benefits may depend on certain aspects of current technology y_{bt}. Technological innovation may increase or decrease $B(p_t)$ by finding new uses or substitutes for GHGs.

293

$f_t(\theta; y_{\theta t})$, subjective probability distribution function for θ. It depends on current knowledge about the relationships between GHGs, global warming, solar radiation, and effects on the biosphere $y_{\theta t}$. We assume scientists and policy makers can agree on a common probability function to characterize current understanding of these relationships. $f_t(\theta; y_{\theta t})$ is revised in each period in accordance with Bayes' rule, reflecting any knowledge gained during the period. (Gottinger, 1980) For example, if the value of some outcome variable z_t is observed and the distribution for z_t conditional on θ is known to be $g(z_t \mid \theta)$, then $f_{t+1}(\theta \mid z_t) = (f_t(\theta) g(z_t \mid \theta)/ \int f_t(\theta) g(z_t \mid \theta) d\theta$. In general, the rate at which knowledge about the chemical, physical, and biological relationships between GHG emissions and effects on life and important materials increases will be a stochastic function of cumulative research expenditures. It may also depend on past and current GHG emissions, since if emissions are greater, any effects on the atmosphere will be larger, detectable effects will occur sooner, and so knowledge about them will be developed more rapidly. In the opposite case, if GHG emissions are eliminated we may never learn whether significant global warming would have occurred.

ϕ_t, current investment in technology and knowledge. ϕ_t is the sum of current expenditures on research addressing each area of technology or knowledge, ϕ_{bt}, $\phi_{\theta t}$, ϕ_{et}, and ϕ_{dt}. Knowledge and technology in specific areas y_{bt}, $y_{\theta t}$, y_{et}, and y_{dt}. are all cumulative functions of past investments, for example, $y_{bt} = \Sigma_l \phi_{bt-l}$. (We assume no decay of knowledge and technology, although this could be readily incorporated.) Relationships describing the productivity of research and development expenditures ϕ on D, B, e and $f(\theta)$ are implicit in the functions D, B, e, and the updating rule for $f(\theta)$.

State transition function:

$$s_{t+1} = s_t + a_s(0) e_t - \Sigma_l \ [a_s(l) - a_s(l+1)] \ e_{t-l}$$

If s_t can be represented as a weighted sum of past GHG emissions, the new state depends on decomposition of the old atmospheric burden plus current GHG emissions, as specified.

Value function:

$$V_t(s_t, y_t, p_t) = B(p_t, y_{bt}) - E[f_t \ D(s_t, y_{dt}, \theta)] - \phi_t$$
$$+ \delta \ E[f_t \ V^*_{t+1}(s_{t+1}, y_{t+1})]$$

where $V^*_t(s_t, y_t) = \text{Max}_p [V_t(s_t, y_t, p_t)]$ and δ is the one-period discount factor. (The value function is also known as the fundamental recursive relation or the Bellman equation.)

The policy problem is to maximize $V_t(s_t, y_t, p_t)$ in the current period. The decision variables are current GHG production and use (p_t) and current expenditures (y_t) on research and development (ϕ_t), allocated among projects to refine estimates of the causes and consequences of global warming ($\phi_{\theta t}$), damage mitigation strategies (ϕ_{dt}), GHG substitutes (ϕ_{bt}), and emission reducing measures (ϕ_{et}). Not all research and development expenditures will be funded out of public resources, but to account for changes in social welfare we must include the real social cost of these projects regardless of the direct funding agency.

At each state, current damage $D(s_t, y_{dt}, \theta)$ is observed. Over the near future, this observable damage is expected to be zero, or nearly zero. At present, it is not clear whether we can distinguish any greenhouse warming that may have occurred from natural fluctuations. D is defined as including only currently observable damage. The fact that future damage is expected to depend on current GHG emission is reflected in the dependence of D on s_t, that is, on cumulative emissions, and thereby on past GHG concentrations.

Observation of D provides information about θ, thereby shifting $f_t(\theta; y_{\theta t})$, but uncertainty about θ remains.

The horizon is indefinite. It may be infinite, or effectively so, if the possibility of greenhouse warming poses a permanent constraint on human activities. In this case, in the very long run, optimal emissions may be equal to some constant, non-zero equilibrium level at which GHG concentrations remain constant. The GHG carrying capacity of the geosphere may be dependent on the chosen equilibrium GHG concentration. However, at present, it is not clear whether the atmosphere GHG concentration is dynamically stable above certain GHG emission levels. If the long-run solution is a constant non-zero GHG emission level, the question of how quickly this equilibrium is approached is a policy question that involves balancing the costs of emission reductions against those of transient and possible permanent environmental damage.

The formulation of the problem suggests that one might change the policy affecting GHG use and emissions in every period, tightening and relaxing emission controls as appropriate. A number of factors limit the desirability of this type of optimization, however, and complicate solution of the dynamic program. For example, the benefits and damages associated with GHG use and the atmospheric response to GHG emissions may exhibit discontinuities, non-convexities, indivisibilities, and irreversibilities. If a policy to reduce GHG emissions induces an industry to close down, subsequent relaxation of the policy may not lead the industry to redevelop. It may be that the industry was profitable before it was shut down, because the non-salvageable component of investment in physical and human capital had already been sunk, but the industry is not sufficiently profitable to induce investors to re-invest in these non-salvageable components when the emissions policy is relaxed. Analogously, requiring firms to install modest emission-limiting equipment may be inefficient if there is a substantial chance that more effective equipment may be necessary later, assuming the first set of equipment cannot be salvaged or upgraded if necessary. In this case, the costs sunk in the first set of equipment would be lost when the second set is required.

The perception that regulations will be subject to frequent revision could also provide a disincentive to invest in physical or human capital that is not fully salvageable, since investors have little assurance of being able to recover their costs before new emission restrictions limit their output or increase production costs. As a result, existing industries may fail to invest in new, more efficient technologies and consequently operate above their long-run cost curves. Similarly, potentially profitable, social-welfare increasing industries may not become established.

In this most general form, the problem is mathematically intractable since it combines stochastic dynamic programming with endogenous learning. However, it is possible to take advantage of specific features of the problem to simplify it in such a way as to provide insight into the current policy question - whether to adopt additional regulations to limit GHG use and emissions now, or delay further regulation until we have a better understanding of whether the likely magnitude and consequences of global warming make it imperative.

To simplify the dynamic program, consider the value function currently facing policy makers, where the current time is denoted 0.

$$V_0(s_0, y_0, p_0) = B(p_0, y_{b0}) - E[f_0 D(s_0, y_{d0}, \theta)] - \phi_0 + \delta E[f_0 V_1^*(s_1, y_1)]$$

Assume that $E[f_0 D(s_0, y_0, \theta)] \approx 0$, since it is not clear whether significant GHG-induced global warming has occurred and no resulting damages have been identified. Significant damages are not expected to occur for some time, since many of the potential damages from GHG accumulation are delayed.

Second, assume that the choice of ϕ_0 can be neglected, either because the range from which it can be selected is small relative to the other terms in the value function, or because the research budget is set exogenously and is not subject to policy makers' influence or control, Typically, a large share of current research is conducted by industry; government policy makers may have little influence over the extent and direction of this work.

Using these assumptions, we can simplify the value function to:

$$V_0(s_0, p_0) = B(p_0) + \delta E[f_0 V_1^*(s_1)]$$

where the notation showing the dependence on y has been suppressed because technological and scientific development is taken as exogenous. For specificity, assume that smaller values of θ correspond to states for which greenhouse warming is more likely or harmful. In what follows, we drop the time-subscript 0 to simplify notation; the optimal current action is denoted p*.

This formulation captures the trade-off between current benefits from emitting GHGs, B(p), and potential future damages, incorporated in V*(s$_1$). Since benefits accrue in the current period but damages will not be incurred until some time in the future, the discount factor δ is a key parameter in the analysis. A higher discount factor (or lower discount rate r, where $\delta = [1 + r]^{-1}$), representing greater concern about future conditions relative to the present, will decrease optimal current production p*. A lower factor (corresponding to a high discount rate) assigns relatively less weight to future benefits and damages, and so increases p^*.

This formulation also provides insight into the role of uncertainty and learning. If the primary source of uncertainty is whether or not GHG emissions will lead to global warming, but the consequences of such global warming, should it occur, are comparatively certain, then greater uncertainty in the sense of a smaller probability that global warming will occur increases the optimal p. In this case, the current benefits of GHG emission are relatively certain but the possible damages are contingent on global warming actually occurring. The less certain policy makers are that such warming will occur, the smaller are the expected damages, and thus the smaller the current benefits that should be foregone to prevent these damages. If, over time, scientific research increases the perceived likelihood of global warming occurring, the optimal p_t will decline.

In contrast, a reduction in uncertainty about the likelihood and consequences of global warming, holding the expected damages constant, may or may not affect the desirability of adopting additional current emission restrictions. The trade-off explicit in the value function is between known current benefits and the expected value of future net benefits (benefits and damages). If the expected utility of future net benefits is held constant, the desirability of current regulations is unaffected by the degree of uncertainty. However, if the expected damages, as measured in money, are held constant and social utility is a non-linear function of the monetized damages over this range, uncertainty may affect the trade-off. In the more likely case, society may be risk averse with regard to these damages, so greater uncertainty about the net benefits will be reflected in a larger risk premium and will enhance the attractiveness of additional current emission restrictions. Decreasing uncertainty over time will reduce the risk premium and thus increase the optimal p_t.

The set of possible decisions is effectively constrained, since $p \geq 0$ and the maximum feasible p is limited by demand for the products and processes to which GHGs are suited. If the density function $f(\theta)$, benefit function $B(p)$, and optimal value function $V_1^*(s_1)$ are reasonably well-behaved, the optimal p^* is likely to be a monotonic function of θ. That is, if $f^a(\theta)$ stochastically dominates $f^b(\theta)$ (simplistically, it assigns greater probability to larger values of θ) the corresponding optimal production is $p^a > p^b$.

Assume that, for political or other reasons, the choice of decision is constrained so that only two policies, p^- and p^+, can be considered. For example, p^- could represent

298

adopting a specified set of proposed emission restrictions that would become effective immediately and p^+ could represent maintaining the existing regulations while delaying one period to await the results of a scientific assessment study. Then, p^- will be optimal for all $\theta < \theta^*$ and p^+ will be optimal for $\theta > \theta^*$. Similarly, if the value of θ is unknown but its distribution is $f(\theta)$, p^- will be optimal for some distributions $\{f^-(\theta)\}$ and p^+ will be optimal for the remaining set of distributions $\{f^{+}(\theta)\}$, the first will assign relatively large probability to small values of θ, and elements of $\{f^{+}(\theta)\}$ will assign relatively large probability to large θ values.

If the set of possible distributions $\{f(\theta)\}$ is suitably limited, it is possible to describe the boundary between the sets $\{f^-(\theta)\}$ and $\{f^+(\theta)\}$ by simple constraints on their parameter values. For example, if $\{f(\theta)\}$ includes only triangular distributions on an interval $[\theta^-,\theta^+]$, $\{f^-(\theta)\}$ will include only those distributions for which $E(\theta) \mid f(\theta) < \theta'$ for some value θ, and $\{f^{+}(\theta)\}$ will include all distributions with $E(\theta) \mid f(\theta) > \theta'$. If $\{f(\theta)\}$ includes a more general set of distributions, the boundary between the sets $\{f^-(\theta)\}$ and $\{f^+(\theta)\}$ may involve higher moments of the distribution as well as the expected value.

This simplification allows us to represent the current policy problem as the simplified stage of the dynamic programme shown in Figure 2. The current state, incorporating past GHG emissions, is s. The possible decisions for the current period are p^-, representing new, additional GHG emission restrictions, and p^+, representing no additional regulations in this period. Regardless of the policy chosen, we assume that scientific research during the period will allow us to revise our current probability distribution function from $f(\theta)$ to one of two possible distributions, $f_1^+(\theta)$ or $f_1^-(\theta)$. The first corresponds to learning that potential greenhouse warming is not as serious a problem as we fear, so the expected value of future benefits and damages conditional on optimal policy thereafter, denoted V^{*1} or V^{*3}, is relatively high. The corresponding optimal decisions in the next period, p_1^1 or p_1^3, depend on whether p^+ or p^- was chosen in the current period. If p^- was selected (additional regulations were imposed), the optimal policy p_1^3 might require relaxing these restrictions; if p^+ was selected (no additional restrictions in the current period), p_1^1 might correspond to continuing the policy of maintaining pre-existing regulations while monitoring scientific developments.

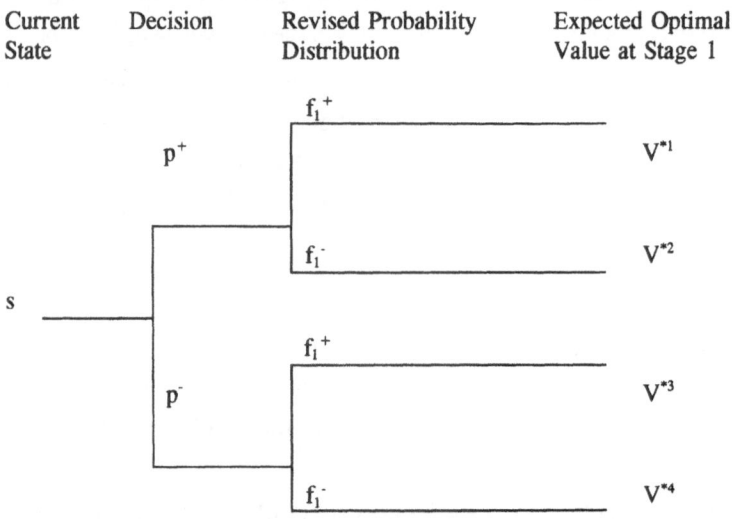

Current State	Decision	Revised Probability Distribution	Expected Optimal Value at Stage 1

Figure 2: Simplified Current Stage of the Dynamic Programme

The second distribution, $f_1^-(\theta)$, corresponds to learning that greenhouse warming is more likely to have serious adverse consequences. The corresponding expected values, V^{*2} or V^{*4}, are comparatively small; $V^{*2} < V^{*1}$ and $V^{*4} < V^{*3}$. The optimal policies in the subsequent period, p_1^2 or p_1^4, correspond to imposing further restrictions to limit GHG emissions. Because emissions in the current period were not limited along the path leading to V^{*2}, the optimal policy in the next period, p_1^2, involves more stringent additional emission limitations than the policy p_1^4 that is optimal if additional emission restrictions were adopted in the current period, so $V^{*2} < V^{*4}$.

9.4 OPTIMAL POLICIES IN A STOCHASTIC DYNAMIC PROGRAMME

The optimal decision in the current period is the value of p that maximizes the value function

$$V(s,p) = B(p) + \delta E [fV_1^*(s_1)]$$

Because we assume that the revised distribution $f_1(\theta)$ can be one of only two possible distributions, $V_1^*(s_1)$ is a Bernoulli random variable and $E[fV_1^*(s_1)]$ depends on the subjective probability α that $f_1(\theta) = f_1^-(\theta)$. The expected values

of the two candidate policies are:

$$V(s,p^+) = B(p^+) + \delta [(1 - \alpha) V^{*1} + \alpha V^{*2}]$$
$$= (1 - \alpha) V^1 + \alpha V^2$$

$$V(s,p^-) = B(p^-) + \delta [(1 - \alpha) V^{*3} + \alpha V^{*4}]$$
$$= (1 - \alpha) V^2 + \alpha V^4$$

where $V^i = B(p^+) + \delta V^{*i}$ for $i = 1,2,3,4$. V^1, V^2, V^3, V^4 represent the value of each path at the current stage, assuming optimal decisions at all subsequent stages.

As illustrated in Figure 3, the optimal choice between the policies depends critically on α. The figure shows the expected benefits $V(s,p) = B((p) + \delta E [f V_1^* (s_1)]$ for each policy as a function of α. For small α, $V(s,p^+) > V(s,p^-)$, since small α indicates that greenhouse warming is not likely to be a serious problem, so significant emission restrictions are not likely to be required in future periods. For large α, $V(s,p^+) < V(s,p^-)$ since further restrictions are likely in future periods. The value of α for which the expected values conditional on each policy are equal is called the "critical probability" q.

Figure 3 also illustrates the expected cost of choosing the wrong policy. The expected cost is a linear function of the difference between the subjective probability that emission-limiting regulations will be necessary and the critical probability. When $\alpha = q$ the expected cost is zero; for other values of α it can be calculated as:

$$E(L) = |B(p^+) - B(p^-) + \delta [(1 - \alpha) (V^{*1} - V^{*3}) + \alpha (V^{*2} - V^{*4})]|$$

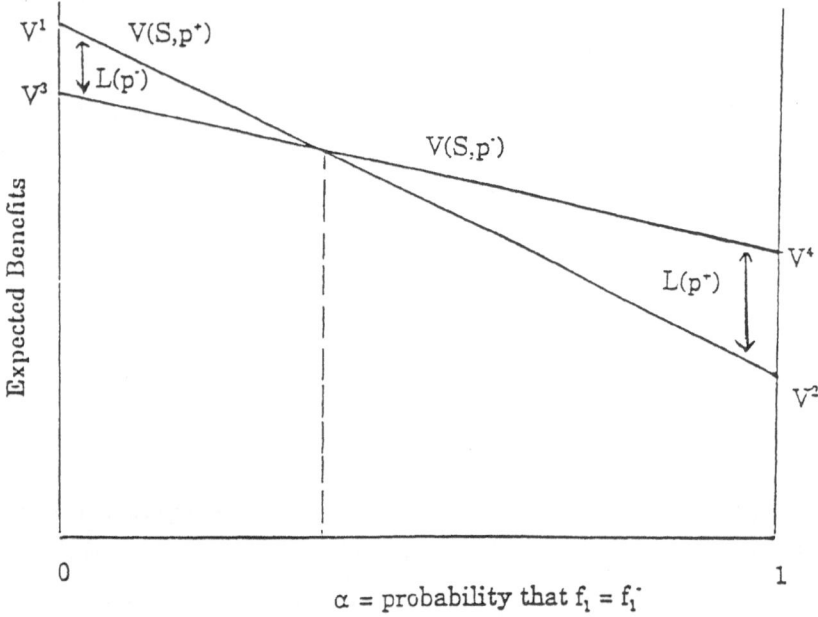

Figure 3: **Expected Benefits of Alternative Current Policies. The lines labelled V(s,p⁺) and V(s,p⁻) show the expected benefits of policies p⁻ (impose additional restrictions on GHG emissions) and p⁺ (do not impose additional restrictions at present) as a function of the probability α that additional emission restrictions will be required in the near future.**

The costs of choosing the policy that is revealed with hindsight to have been incorrect, $L(p^+)$ or $L(p^-)$, are:

$$L(p^+) = - [B(p^+) - B(p^-) + \delta (V^{*2} - V^{*4})]$$
$$= V^4 - V^2$$

$$L(p^-) = - [B(p^-) - B(p^+) + \delta (V^{*3} - V^{*1})]$$
$$= V^1 - V^3$$

$L(p^+)$ measures the loss if p^+ (no new regulations) is chosen in the current period but we learn that substantial emission reductions are necessary ($f_1(\theta) = f_1^-(\theta)$). Since $B(p^+) >$

302

B(p⁻), the higher benefits from choosing p⁺ in the current period partially offset the higher damages and control costs incorporated in (V^{*2} - V^{*4}). Analogously, if p⁻ (additional emission restrictions) is chosen but we learn these restrictions are not necessary ($f_1(\theta) = f_1^+(\theta)$) the cost is L(p⁻). The foregone benefits in the current period, B(p⁻) - B(p⁺), may be partially offset by possible increases in future production and emissions reflected in (V^{*3} - V^{*1}). L(p⁺) largely reflects the long-term control-cost savings possible if emission restrictions begin earlier; L(p⁻) primarily reflects foregone benefits in the current period due to unnecessary restrictions.

The critical probability solves the formula:

$$q = \cfrac{1}{1+\cfrac{L(p^+)}{L(p^-)}} \qquad L(p^-)>0, L(p^+)>0$$

$$\left. \begin{array}{ll} q = 0 & L(p^-) \leq 0 < L(p^+) \\ q = 1 & L(p^+) \leq 0 < L(p^-) \end{array} \right\}$$

It depends on the relative costs of the two possible errors. If either possible cost is not positive the critical probability degenerates to zero or one, and one of the decisions is always better. In the usual case, however, both losses will be positive and q is a function of their ratio. If the ratio L(p⁺)/L(p⁻) is large, the cost of delaying additional emission restrictions when such restrictions will be necessary in later periods overwhelms the cost of imposing additional restrictions that later prove unnecessary. Thus, q is small, and p⁺ is preferred only if policy makers are very sure that subsequent emission restrictions will not be needed. Alternatively, if L(p⁺)/L(p⁻) is small, the costs of stringently restricting emissions in the future are small relative to the foregone benefits of limiting emissions in the current period, q is large, and additional regulations should not be imposed in the current period (p⁻) unless policy makers are reasonably certain that emission reductions will be necessary in the future.

Evaluating the critical probability requires estimating the values of the losses from choosing each policy when, after the fact, the other policy turns out to have been preferred, L(p⁺) and L(p⁻). These losses in turn depend on the expected values assuming optimal regulations thereafter, V^{*1}, V^{*2}, V^{*3}, and V^{*4}, Thus, the solution to the current policy question depends on the expected welfare conditional on optimal regulations in each of the following periods.

Since stochastic dynamic program require knowing the expected value of future stages to determine optimal decisions, there are two solution methods. (Ross, 1983, Chapters 1,2) If the problem has a finite horizon, one can begin with the final period. The expected values corresponding to each possible terminal state are calculated. From these, one can find the optimal expected values and policies corresponding to each possible state at the penultimate stage. Using the optimal values for each penultimate state, one can solve for the optimal expected values in the proceeding stage, and continue back to the current stage.

If the dynamic programme does not have a finite horizon, as in this case, solution is not straightforward. Often, such problems can be solved only if the optimal values and policies are constant or geometrically declining over time. Solution may depend on being able to write the value function and state transition function in simple analytic form.

In order to describe the solution to the current stage of the policy problem illustrated in Figure 4, we make a series of assumptions that enable us to approximate the optimal values V^{*1}, V^{*2}, V^{*3}, and V^{*4} in the next period. Specifically, we assume that the state s can be adequately represented by cumulative weighted GHG emissions, where the weights correspond to the estimated relative global warming efficiencies of each compound.

Second, we assume that optimal policies beginning in the next period will produce equal cumulative weighted emissions as of a fixed future date, denoted the planning horizon. Regardless of which policy is chosen for the current period, no significant global warming is likely to occur before the horizon so the damages associated with warming are negligible and do not depend on the current policy choice. Thus, the only difference between the expected optimal values in the next stage depend on the costs of reducing future GHG emissions: If emissions are not restricted in the current period, achieving the same cumulative emissions through the horizon will require more stringent restrictions in future periods. Since the marginal cost of reductions is presumably increasing and convex, the present value of the cost of achieving a fixed cumulative decrease in emissions through the horizon may increase if the restrictions are constrained to a shorter time period.

A more precise solution would recognize that the optimal level of damages to accept might be greater if GHG emissions are not restricted in the current period. Since the cost of achieving a fixed reduction in cumulative emissions through the horizon may increase if current emissions are not restricted, it might be optimal to accept slightly more warming-induced damage, and incur slightly smaller control costs in this case. However, the error introduced through this assumption should not be large.

Third, to calculate the optimal expected values corresponding to each of the four possible states s_1, we assume the expected value is the same as the value if the optimal cumulative weighted GHG emissions through the horizon became known at that time. Thus, f_1^- and f_1^+ might each concentrate their mass at a single point, so that at the next stage we know that cumulative emissions should be limited to ψ^- (if the new distribution is f_1^-) or ψ^+ (if it is f_1^+). One way to understand this assumption is to assume that, at the end of the current period, scientific research will yield sufficient understanding of the causes and consequences of warming that we can then select an optimal level of GHG emissions, balancing the costs of warming against those of emission control. The optimal cumulative emission level will be either ψ^- or ψ^+. A more general interpretation would be that uncertainty about the appropriate level of cumulative emissions through the horizon will be substantially reduced, and that ψ^- or ψ^+ reasonably characterize the possible cumulative emission limits that will appear to be appropriate at that time.

This simplified analysis abstracts from several important features of the problem in order to make it possible to calculate a numerical solution and provide quantitative guidance to policy makers.

First, it assumes that information will arrive early enough that significant adverse effects can be prevented by imposing sufficiently stringent regulations at that time. The only penalty for not imposing additional emission restrictions in the current period is the lost opportunity to distribute any required emission reduction over a longer period, and thereby potentially reduce their cost. Since the environmental consequences do not depend on whether regulations are adopted immediately or not, the analysis reduces to a comparison of the expected economic costs of the alternatives.

Second, many aspects of the simplified problem are discrete, not continuous as in the real problem. For example, in the simplified problem only two information outcomes are possible: a more realistic subjective distribution for the appropriate level of emission limits would assign positive probability to a continuous range of emission limits. To offset this limitation, the critical probability can be calculated for the entire range of possible cumulative-emission limits, to assess its sensitivity to these choices. Similarly, new information that substantially reduces uncertainty about the ultimate consequences of emissions arrives in a discrete package at a predetermined date. This simplification should bias the results in favour of delayed contingent regulations, because in the real problem information will arrive in smaller bits at irregular, somewhat unpredictable intervals and the uncertainty will never be completely resolved, so we can never avoid the risk of regulating more stringently than necessary.

Third, in the simplified problem the date and type of new information are independent of the chosen regulatory strategy (learning is passive). In fact, the rate of scientific progress may depend on the regulations chosen: in the extreme case, if emissions are severely limited we might never learn whether significant global warming would have occurred.

Finally, the simplified problem abstracts from the issue of choosing the appropriate level of emissions and corresponding environmental and welfare consequences. In principle, this issue can be solved by comparing the costs of environmental modification with those of emission control, although the consequences and costs of environmental change cannot credibly be measured at present. Despite these simplifications, this simplified problem elucidates many of the issues pertinent to the decision.

9.5 RESOURCE COSTS AND CRITICAL PROBABILITIES

In this section we apply the previous approach through the calculation of the optimal values, and make use of simulation of demand curves for GHG related products and processes based on analysis of the costs of substitute and manufacturing processes (Cline, 1992). We show that this method works for timing regulations depending on the critical ("threshold") probabilities.

Calculation of the critical probability for any proposed immediate regulation and cumulative emission limit requires calculation of the present value of the resource costs of alternative regulatory trajectories (the values V^1, V^2, V^3 and V^4 in Figure 3). These resource costs are measured as areas under derived demand curves for GHGs.

The calculations assume that the regulations will consist of taxes imposed on GHGs. Using a tax, the effective price of GHG related products can be increased so that consumers will switch to alternate products and manufacturers will substitute other more conservative processes, or other technological options that become cost effective. The size of the tax can be varied to induce the desired amount of emission reductions. The use of such a tax induces manufacturers and consumers to adopt the economically efficient set of emission reducing measures, thereby minimizing the annual resource costs of emission reductions. To minimize the present value of the cost of limiting cumulative weighted emissions, the taxes should be proportional to the relative GHG contributing share of each product or process and rise over time at the discount rate that firms and consumers use in making investment and consumption decisions.

If a tax is applied, the resource costs of the regulation can be measured by the area under the derived demand curve for each GHG between the unregulated price and the price including tax.

The level to which cumulative weighted emissions can be constrained depends on the date at which emission regulations are imposed and the stringency of the regulations. Figure 4 illustrates the effect of these factors. The abscissa indicates the initial base tax, ranging between zero and five monetary units (MUs). In the standard case, GHG demand is assumed to grow and the tax increases 3 per cent per year. The three lines in the figure correspond to regulations beginning in 1995, 2010, and 2030. Suppose, in the absence of regulations (that is, with a tax equal to zero), global cumulative weighted emissions from 1995 through 2030 total about 63.5 billion metric tons (bMt). If regulations were to begin in 1995, limiting emissions to 50 bMt would require an initial world-wide tax of about MU 0.90/lb, limiting emissions to 40 Mt would require an initial tax of about MU 1.90/lb, and the minimum attainable level of cumulative emissions, if the initial tax were MU 5/Mt would be about 32.5bMt. If regulations were not initiated until 2010 larger taxes would be necessary to limit emissions to the same levels: A 50 bMt limit would require a tax beginning at MU 1.20/Mt; a 40b Mt limit would require a tax beginning at MU 2.83/Mt. The smallest attainable cumulative emissions, if regulations did not begin until 2010, would be about 37.2 bMt. If regulations were not imposed until 2030, the range of attainable cumulative emissions is further reduced, and even higher surcharges would be required to hold emissions to any attainable level.

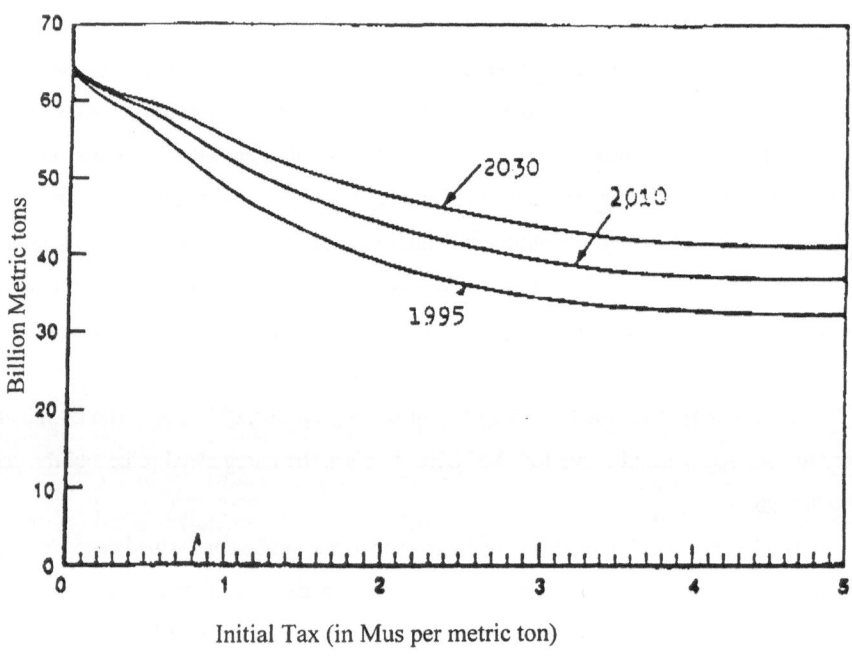

Figure 3: Cumulative weighted emissions through 2030 as a function of the initial data and tax.

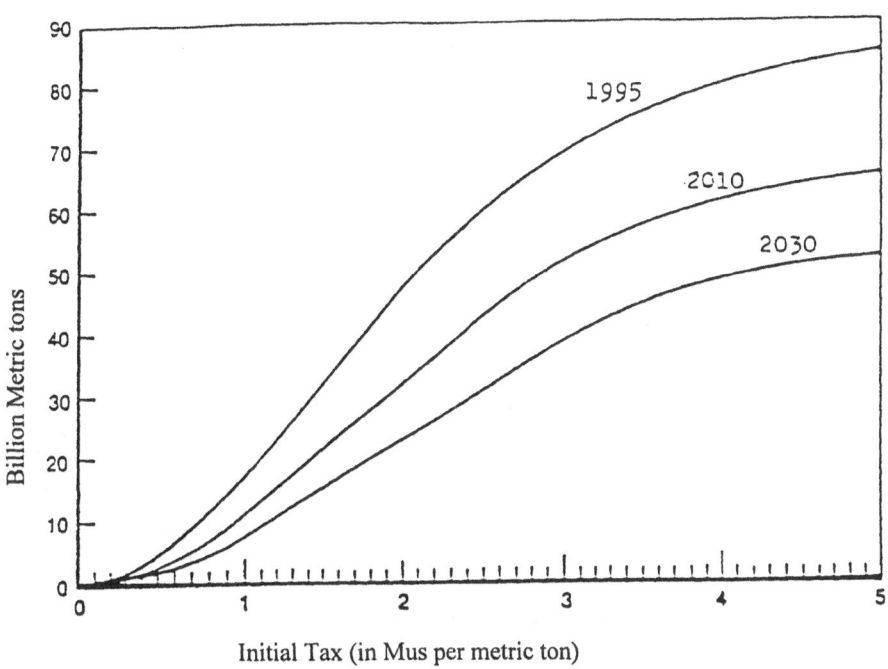

Figure 4:   Present value of resource cost as a function of the initial data and tax, using a 3 % discount rate

Figure 4 illustrates the case of relatively high inelasticity of demand for GHG-related products: Even with taxes of several MUs per Mt, compared with unregulated prices on the order of 0.50 MUs/Mt simulated cumulative emissions over the next 35 years fall by no more than half. Limiting cumulative emissions to 35.2 bMt, the level corresponding to continued emission at 1992 rates would require an initial tax of MU 2.80/Mt, if regulations begin in

1995 and would not be possible (under these assumptions) if regulations were delayed until 2010.

The resource costs (lost economic surplus) associated with restrictions that reduce GHG emissions are substantial. Figure 5 illustrates the present value (in 1992 using a three-percent discount rate) of the resource costs associated with a tax beginning at the level indicated on the abscissa and increasing at three per cent per year. Using Figures 4 and 5, one can estimate the resource cost associated with various cumulative-weighted emission levels. For example, to limit cumulative emissions through 2030 to 50 bMt requires a tax beginning at MU 0.90/Mt if regulations begin in 1995. As shown by Figure 5, the present value of the associated resource cost is MU 14.4 billion. If the regulations are not implemented until 2010 the required initial tax is MU 1.22/Mt, which is associated with a resource cost of MU 15.8 billion in present value.

Figure 6 illustrates the critical probability that determines whether the expected cost of regulating immediately is greater or smaller than the expected cost of waiting for new information before regulating. The critical probability varies with the cumulative emissions that can be tolerated and with the proposed level of immediate regulations. For example, assume that if research during the current period produces f_1^- as the distribution of effects, optimal regulations will limit cumulative emissions through 2030 to 55bMt. If the distribution is f_1^+, the 63.5 bMt of emissions that would occur without regulations will be acceptable. The proposed immediate regulations consist of a base tax of MU 0.30/Mt, increasing 3 per cent per year. If the new information indicates that cumulative emissions must be limited to 55bMt, the tax will be doubled in 2010 (the path leading to V^4 in Figure 2). If the new information indicates that no regulations are required, the tax will be dropped (path 3), and no further costs will be incurred. Alternatively, if the proposed interim regulations are not adopted and the new information indicates that cumulative emissions must be limited to 55bMt, it would be necessary to impose a tax of MU 0.79/Mt in 2010 that would increase 3 per cent annually (path 2). If the new information indicates cumulative emissions of 63.5bMt can be tolerated, no regulations would ever be imposed (path 1).

Projected globally averaged GHG accumulation in 2030 (Per Cent)

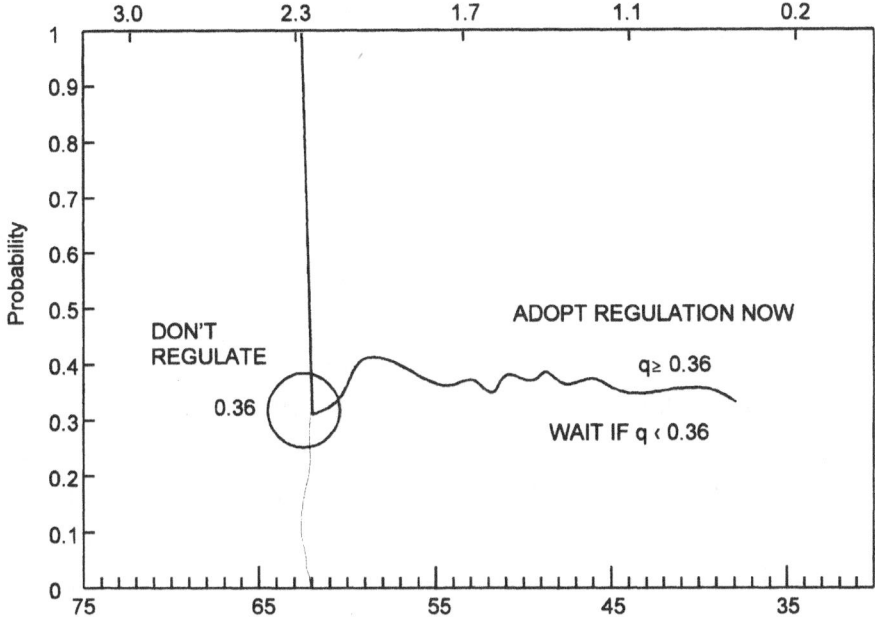

Figure 5: Critical probability as a function of acceptable cumulative weighted emissions through 2030: standard case.

The critical probability can be calculated from the present values of the resource costs associated with each of these four possibilities. Since we do not explicitly evaluate the benefits of GHGs we express these as deviations from the value of the path with no emission restrictions, $V^1 = B(p^+) + V^{*1}$. The values V^1, V^2, V^3, and V^4 are V^1 - 0, V^1 - MU 6,785 million, V^1 - MU 295 million, and V^1 - MU 6,286 million. From these, $L(p^+)$ = MU 6,785 million - MU 6,268 million and $L(p^-) = MU295$ million - MU $\varnothing$. Using the formulae for the critical probability in Section 9.4.

$$q = \cfrac{1}{1 + \cfrac{6785 - 6268}{295 - 0}}$$
$$= 0.36.$$

Thus, if the probability that regulations limiting cumulative emissions to 55bMt will be required is greater that 0.36, the expected cost of adopting regulations now will be less than the expected cost of awaiting new information and regulating in 2010 only if necessary. If the probability that such regulations will be required is less than 0.36, the expected costs of waiting will be lower than the expected costs of regulating now.

The critical probability varies with the level of cumulative emissions that can be tolerated, but is essentially dichotomous. As shown in Figure 4, if the level of cumulative emissions that can be tolerated is approximately equal to the unregulated level (63.5bMt) or higher, immediate regulations cannot be cost justified. However, if emission reductions may be necessary (to lower cumulative emissions to 62bMt or less), the critical probability drops to about 0.35. Over the cumulative emission range from about 62 to 38bMt the critical probability is nearly constant. Finally, if the new information might indicate that emissions must be limited to 37bMt or less, the critical probability cannot be calculated since it is not possible to limit emissions to this level using the assumed demand curves if regulations are delayed to 2010. Thus, Figure 4 indicates that, if the level of acceptable cumulative emissions equals or exceeds the unregulated level, immediate regulations are not appropriate. If some emission reductions may be required (about 25bMt or less), it is cost effective to wait for better information only if the probability that such reductions will be necessary is less than about 0.35. If larger reductions may be necessary, the critical probability cannot be calculated using the simulated demand curves.

For reference, the abscissa of Figure 4 also indicates the projected decrease in globally averaged GHG accumulation in 2030 corresponding to various levels of cumulative emissions (indicated along the top of the figure).

312

Although one might expect the critical probability to decline with the stringency of the proposed immediate regulations, this is not necessarily the case. Recall from the definition of the critical probability that it depends on the ratio of the costs of choosing each policy if it proves wrong ex post, $L(p^+)/L(p^-)$. $L(p^+)$ largely reflects long term cost savings from beginning emission restrictions earlier; $L(p^-)$ primarily reflects foregone near term benefits because of unnecessary restrictions.

9.6   CONCLUSIONS

Formulation of sensible policies for dealing with global warming is greatly complicated by some fundamental scientific uncertainties that are unlikely to be fully resolved in the near future.   This poses an awkward policy dilemma: by the time reliable answers are forthcoming, the damage inflicted on us will have increased greatly, if the pessimists turn out to have been right and we do not follow their prescriptions for drastic changes now. On the other hand, enacting a drastic programme may impose economic costs and social disruptions, especially on poorer regions, that would be clearly excessive if the problem turns out to be less severe than many currently anticipate.

Over the past ten years, with increasing intensity and persistence, there have been a number of major reduction plans of GHG emissions or its components. Proponents of such plans have been arguing that significant reductions are justified as some kind of insurance policy against long term climatic change and disruption.   Others suggest that before embarking on costly programmes we should first remove scientific uncertainties associated with the physical mechanisms likely to induce global warming.   Moreover, if such mechanisms were shown only to cause moderate warming one may rely more on adaptive, market driven strategies than on complex, costly regulatory schemes. (Gottinger and Barnes, 1993, Schelling, 1988)

Our model allows calculation of a 'critical probability' that characterizes the conditions under which the insurance benefits of immediate regulations exceed their cost. This critical probability can function like the standard of proof required in judicial settings. Like the standard of proof, it specifies the degree of confidence policy makers must have that GHG emissions will need to be restricted to avoid significant adverse ecological effects in order for them to judge that additional emission restrictions should be adopted at present. If policy makers' perceived probability that emission reductions will be required is greater than the critical probability, the strategy of adopting regulations immediately will impose lower expected resource costs; of the probability is lower, waiting for improved understanding of the likelihood and consequences of GHG accumulation before acting will be cost effective.

The conventional decision analysis approach would require policy makers to assess their complete subjective probability distributions for the extent and consequences of the future. This distribution would be used to integrate the value of alternative outcomes across branches of a decision tree, and the output would consist of expected values corresponding to

314

alternative policies. With this approach, the role of the subjective probability judgments and the sensitivity of the policy choice to variations in these judgments would be concealed through the integration. Sensitivity analysis would require recalculating the expected values for each probability distribution.

In contrast, the critical probability approach focuses attention on the subjective probability judgement and makes its role in the conclusions transparent. This approach does not require scientists an policy makers to develop a complete distribution, but only to assess whether the bulk of the distribution lies to one side or the other of a specified cut off. It reduces the level of agreement needed to obtain a consensus for policy and clarifies the beliefs that require agreement. Policy makers may be reluctant to be publicly identified with a precise distribution but more willing to state whether they believe the evidence is sufficiently convincing or not. Because assessing a complete subjective probability distribution is time consuming, unfamiliar, and not well done by many policy makers (Kahneman et al., 1982), this alternative to the conventional approach may be of value.

The application presented here is particularly simple, because we have considered only Bernoulli probability distributions, for which a single probability characterizes the entire distribution. But the approach could be extended to more general distribution. If it were, policy makers might have to make a more difficult judgement than one about the level of a single probability, but they would only have to determine whether their own distribution correspond to one or another subset of the admissible distributions. In many cases, this should be an easier task than specifying the shape of the entire distribution.

Sensitivity analysis for his problem shows that, over a wide range of assumptions, the critical probability is a nearly dichotomous function of the extent of emission reductions that may be necessary. If the cumulative emissions that will occur in the absence of additional regulations will not produce significant adverse environmental changes, immediate regulations cannot be cost effective. If emission reductions may be necessary, the critical probability falls between about 0.3 and 0.5 over the domain of cumulative emission limits for which it can be calculated.

GHG emission and climatic change are global issues: Their effects, if realized, will be felt world wide. This paper explicitly avoids the important issues associated with the coordination of action among nations. It focuses instead on the logically prior question of whether, form a global perspective, immediate regulations may be appropriate. The results

315

suggest that whether immediate regulations are cost justified depends primarily on the quantity of future emissions that is acceptable and the likelihood that regulations to line emissions to that level will be necessary.

REFERENCES

Bertsekas, D. P., *Dynamic Programming and Stochastic Control*, Academic Press: New York 1976.

Conrad, J. M., "Stopping Rules and the Control of Stock Pollutants", *Seminar on Uncertainty in Management of Natural Resources and the Environment*, Central Statistical Bureau, Oslo 1992.

De Finetti, B., *Probability, Induction and Statistics*, Wiley: New York 1972.

Fishburn, P., *The Foundations of Expected Utility*, Reidel: Dordrecht 1982.

Flohn, H., "Treibhauseffekt der Atmosphäre: Neue Fakten und Perspektiven" ("Greenhouse effect of the atmosphere: new facts and perspectives"). *Rheinisch-Westfälische Akademie der Wissenschaften* Vorträge N 379, 1990, pp.9-48.

Gottinger, H. W., and P. Barnes, "Energy, Economy and the CO_2 Problem", Chapter 2-6 L. Hens (ed.), *Hand book of Environmental Management*, Free University of Brussels Press: Brussels, 1993.

Gottinger, H. W., *Elements of Statistical Analysis*, de Gruyter: Berlin, New York, 1980.

Hammitt, J. K., R. J Lempert and M. E. Schlesinger, "A Sequential-Decision Strategy for Abating Climate Change" *Nature* 357, 1992, pp. 315-318.

Henrion, M. and M.G. Morgan, *Uncertainty: Guide to dealing with Uncertainty in Quantitative Risk and Policy Analysis*, Cambridge Univ. Press: Cambridge 1990.

Kahneman, D., P. Slovic, and A. Tversky, *Judgement under Uncertainty: Heuristics and Biases*, Cambridge University Press, 1982.

Kelly, D. L. and C. D. Kolstad, "Tracking the Climate Change Footprint: Stochastic Learning about Climate Change", University of California, Santa Barbara, WP in Economics #3-96R, 1996

Kolstad, C. D., "Looking vs. Leaping: The Timing of CO_2 Control in the Face of Uncertainty and Learning", IIASA Workshop on Costs, Impacts and Possible Benefits of CO_2 Mitigation, IIASA, Laxenberg, Austria, September 1992.

Lave, L. B., "Formulating Greenhouse Policies in a Sea of Uncertainty", *The Energy Journal* 12(1), 1991, pp. 9-21.

Lempert, R. J., Schlesinger, M. E. and J. K. Hammitt, "The Impact of Potential Abrupt Climate Changes on Near-Term Policy Choices", Climatic Change 26, 1994, 351-376.

Manne, A. S. and R. G. Richels, *Buying Greenhouse Insurance*, MIT Press: Cambridge, Mass., 1992.

Parry, I. W. H., "Some Estimates of the Insurance Values against Climate change from Reducing Greenhouse Gas Emissions", *Resource and Energy Economics* 15, 1993, pp. 99-115.

Peck, St. C. and T. J. Teisberg, "Global Warming Uncertainties and the Value of Information: An Analysis using CETA", *Resource and Energy Economics* 15, 1993, pp. 71-97.

Ross, S.M. *Introduction to Stochastic Dynamic Programming*, Academic Press: San Diego, California 1983.

Schelling, T., "Global Environmental Forces", Energy and Environmental Policy Center, Harvard University, E-88-10, November 1988.

Epilogue

The major purpose of this study is to improve economic models for the GHG problem; and to make recommendations for large scale models.

As we covered most of the state-of-the-art EEE-modeling we noticed that many of the models lack integrating features: they are not optimizing, feedback effects are not considered, capital investment is not traced, and uncertainty is not considered. A typical class of models matches demand and supply for energy at discrete points in time in individual world regions. The model structure is relatively simple, but its disaggregation of data both increases realism and makes the results difficult to track.

Energy demand is first projected based on population, economic activity, technological change, energy prices, and energy taxes and tariffs. One class of models determines the supply of primary energy in three different ways. Resource constrained fuels, such as conventional oil and gas, are produced according to resource depletion curves and their supply is unresponsive to price. Hydropower and biomass are considered resource constrained renewable sources which reach and maintain a specified production level. Their production is also insensitive to price. Finally, sources such as coal, nuclear, solar, and unconventional oil and gas are classified as unconstrained. The production levels of these sources are not limited but depend on price.

The primary fuels are converted to secondary fuels and then to energy services. The prices of energy services depend on the price of primary energy, the mix of energy sources providing the service, and transportation and conversion costs. The initial fuel mix is exogenous, but the mix is then modified in light of energy prices.

Conventional model building would move along the chain of events: final energy use is determined by an iterative process which changes prices, balances energy trade among regions and balances energy supply and demand for each region. As energy prices vary, demand and supply adjust in response to exogenously specified demand elasticities and supply functions. Based on the mix and level of fossil fuel use, the carbon dioxide emissions are calculated. The retention of carbon dioxide in the atmosphere and the consequent climate changes are determined by a model which considers other greenhouse gases, ocean absorption of CO_2, and other factors. Another class of models maximizes consumption in individual time periods. A highly aggregate function determines production. It considers no world regions and only two fuel types, carbon and noncarbon. It is not optimizing over time, does not include feedback effects, and does not consider capital stocks. Its most significant characteristic is the estimation of probability distributions for important future variables.

Some modelers assume overall technical progress at a constant rate. The production function relates the energy and labour sectors in a Cobb-Douglas fashion; however, modifications of Cobb-Douglas are common. In the usual Cobb Douglas formulation, the production elasticity of an input is constant and equal to the share of GNP paid to the input. Further, in the usual formulation the elasticity of substitution between inputs is always - 1. These rigidities are a major drawback to the use of the Cobb-Douglas function. For example, in the NRC model (of

Nordhaus-Yohe) the payments to inputs are adjusted in each time period. The adjustment is consistent with a changing, exogenously set elasticity of substitution between labour and energy. One could feel that this overcomes the limitation placed on substitution elasticities by the Cobb-Douglas function. Within the energy sector there is a constant elasticity of substitution between the fossil and nonfossil inputs.

The price of noncarbon based fuel equals the sum of distribution costs and production costs. The distribution costs are constant, but the production costs change exponentially over time. The rate of change is the sum of a term representing the technical change in the energy industry and a term representing a bias toward noncarbon energy.

The equation for carbon fuel prices is similar but somewhat more complex. The first term is again distribution costs. The second term combines both production costs and costs due to fuel depletion. Further the second term is multiplied by the exponential change in energy industry technology, but there is no bias term. Finally, a tax on carbon fuels may be added. The emissions of carbon to the atmosphere per unit of carbon fuel is assumed to grow over time, because the mix of fuels includes more high carbon content fuels. The equation for changes in atmospheric carbon dioxide is similar to that used in Chapter 2 except that, the ratio of emissions to retained atmospheric CO_2 changes over time.

Because the feedback effects of GHGs are extremely uncertain, many modelers are reluctant to incorporate these effects in their models. In some scenarios, feedback effects might indeed be unimportant. For example in models with finite horizons, if GHG effects are insignificant until after the horizon of the model no modeling of feedback effects is needed.

In models which optimize over an infinite horizon, future effects may change current policies, and feedback effects are always of importance. However, in these same models feedback effects may make solution much more difficult. In predictive models with long time horizons, feedback effects will also be important. Further, in predictive models that estimate production and energy use at individual points of time, such as the two classes of models examined here, feedback effects can be easily included.

Evidence through sensitivity analysis shows that in optimizing models inclusion of feedback effects usually lowers the optimal initial use of fossil fuels. The long-run changes in fossil fuel use due to feedback effects are more uncertain and dependent on the model. However, in most models the feedback slows the economy and thus reduces the demand for fossil fuels in the future. If optimization was included in the models examined this effect would be likely.

Given the uncertainty in the severity and timing of feedback effects, the sensitivity of individual models to variations in feedback effects, is of much interest. In Chapter 2 using the step model, we analyzed in a cursory way the sensitivity of current fossil fuel use to an ultimate limit on atmospheric carbon dioxide. We found that current optimal fossil fuel use was significantly affected by different critical levels of CO_2. A study of the impacts of a critical CO_2 level in a more detailed and disaggregated optimizing model would be useful. In this regard a link between an optimizing type model of Chapter 2 and specific growth models of Chapter 7 would be desirable.

Including optimization in models expands their applicability but may cause analytic problems and controversy. As with many social problems, an acceptable objective function for carbon dioxide control problems is difficult to define. Any definition will seem both inadequate and overly precise and certainly will be controversial. This may be the reason some models reviewed did not examine optimal policies; however, statement of an objective function does not hide or confuse other results and can add many new insights. If feedback effects and an objective function are included in a model, a crude optimization can be performed simply by running the model under a variety of policies.

Including optimization raised several new issues in our models. For example, pollution impoverishes but technical progress enriches the future. The optimizing models of Chapters 2 and 3 show how the curvature of the utility function, technical progress and pollution interact. The models show that curvature of the utility function, determined by the consumption elasticity of utility in our models, tends to smooth or even out wealth over time. Without an objective function being stated, the importance of this redistribution effect in determining fossil fuel use policy can not be examined. In predictive models, a subjective evaluation must be made of the significance and value of a policy. In an optimizing model, the costs and benefits of policies are automatically compared in an explicit manner.

A final benefit of an optimizing model is the identification of multifaceted responses which may be ignored when policy changes are specified exogenously. Integrated optimizing models respond to problems by adjusting numerous policies endogenously. For example, in our multiple state models, Chapter 4, fossil fuel use, research, and capital all respond to changes in the effects of GHGs.

Neither of the conventional models examined here include capital. The inclusion of capital could have a variety of effects. The explicit addition of capital would add greater flexibility to the models and thus could improve predictions of the future. Conversely, the inclusion of capital may make certain high consumption growth paths look less attractive. If a high growth requires a high proportion of GNP to be invested, GNP, the simplest measure of welfare, is no longer a good measure of the quality of the growth path. Our models suggest that rapid growth is accompanied by very high investment rates. The inclusion of capital would be quite difficult in either class of models mentioned unless the investment level was exogenously specified.

In vintage models, as in Chapter 4, capital reduces the rate of change in the economy by requiring machine replacement to accompany changes in inputs.

Two structural features of current EEE models deserve comment. Because most of them are not forward looking, the addition of feedback effects and the definition of a welfare function should not be difficult and would provide the opportunity for some simple optimization of policies. Second, the treatment of technical change is unsatisfactory. Technical change is neutral across fuels and the rate of technical change declines over time. Our models suggest, that neutral technical progress depends on special economic circumstances which are not generally expected. The declining rate of technical change is not common in the models examined here.

We feel that two results of environmental studies stand out in particular. The first of these is the importance of other greenhouse gases. There is general agreement that much of the uncertainty regarding future temperature increases is due to these other greenhouse gases, and it is clear that, in part, CO_2 control policies per se are ineffective because they do not change emissions of other greenhouse gases. Our models suggest that the worse the impacts of other greenhouse gases the lower the optimal present emission of CO_2.

The recent IPCC assessment (1996) concludes that it is very difficult to significantly reduce the temperature rise expected in the second half of the next century. This result has two bases: first, the impact of other greenhouse gases and, second, the unresponsiveness of coal use to changing policies. Only a ban on coal instituted by the early 21st century would effectively slow the rate of temperature change and delay the date of a 1.5^0 C change from a base case of the last decades of the 21st century. Worldwide taxes of up to 300% of the cost of fossil fuels, bans on synfuels and shale oil, and other less stringent policies would delay the advent of a $1.5°$ C rise in temperature only by less than 10 years. Price reductions on alternative fuels have little effect in some EEE models because they enhance economic performance and result in higher overall energy demand. This result parallels a result of ours; in Chapter 4 on finite technical progress models, we show that an increase in energy efficiency may increase fossil fuel use.

Other EEE models examine control policies including bans on fossil fuel use, trade restrictions, and fossil fuel taxes. Major models that consider trade restrictions are of particular interest. In a particular run of the models a one country based only energy tax is combined with a limit on energy exports. Such a policy does lower world emissions; however, it should be noted this policy would certainly not be in the best interest of the same country in a competitive world fuel market. In a competitive world market other supplies would replace this country's fuels taken off the market. If the model includes trade, it would be useful to examine the effects of subsidies on nonfossil fuel exports such as fuel conservation equipment, nuclear energy facilities and solar energy facilities. Our model analysis of trade suggests that such subsidies may be worthwhile when international cooperation is incomplete.

In a different context, the addition of feedback effects and the use of an objective function would enhance the models' results. If the model does not link periods through state variables, a simple optimization over multiple policies would be possible. Quite a few models examine the effects of probability but not the particular risks of low periods of consumption. To measure the risk posed by future events we need to know both the probability of the events and how we wish to value periods of high and low consumption. Our models suggest that differences in the valuation of high and low periods of consumption may have a significant impact on present policies. Incorporation of an objective function which considers the curvature of the utility function would allow risk to be examined in a realistic manner.

We also note that inclusion of feedback effects in this model might actually reduce the long-run variance in GNP, because GNP growth and the impacts of CO_2 are negatively correlated. That is, higher GNP growth raises fossil fuel use which may increase the adverse impacts of CO_2 and reduce GNP growth.

In identifying the three most important results in the aforementioned NRC model we note: a close to 30% chance that CO_2 concentration will double before 2050. The possibility of an early doubling of CO_2 makes the incorporation of feedback effects more important. The second is a more than a 10% chance that a nonfossil backstop technology will replace fossil fuels before 2100. The final important result is the statement that the major causes of uncertainty are the ease of substitution between fossil and nonfossil fuels and the rate of technology growth. This final result reemphasizes our concentration on technical change. In this model five different tax policies to control CO_2 were examined. The most stringent of these place a 60% surcharge on the price of fossil fuels. The impact of this stringent policy is to reduce the CO_2 concentrations in 2050 by about 6% as compared to a base case where all parameters are at their median values. Part of the reason for this small change is that the response to such a tax in this model does not include changing technology in capital stocks. For example our flexible technology model in Chapter 4 suggests that a fossil fuel tax would increase research and energy efficiency and cause a relative increase in capital purchases.

Index (numbers in brackets refer to chaps.)

326